Breaking Out of the Green House

Indian Leadership in Times of Environmental Change

Breaking Out of the Green House

Indian Leadership in Times of Environmental Change

Dhanasree Jayaram

Introduction by

Air Commodore **Jasjit Singh** AVSM, VrC, VM (Retd)

KNOWLEDGE WORLD
KW Publishers Pvt Ltd
New Delhi

in association with

Centre for Air Power Studies
New Delhi

2011 FPBA BEST PUBLISHERS AWARD (ENGLISH)

The Centre for Air Power Studies is an independent, non-profit, academic research institution established in 2002 under a registered Trust to undertake and promote policy-related research, study and discussion on the trends and developments in defence and military issues, especially air power and the aerospace arena, for civil and military purposes. Its publications seek to expand and deepen the understanding of defence, military power, air power and aerospace issues without necessarily reflecting the views of any institution or individuals except those of the authors.

Jasjit Singh
Director General
Centre for Air Power Studies
P-284, Arjan Path
Subroto Park
New Delhi 110010

Tele: (91-11) 25699131
E-mail: diroffice@aerospaceindia.org

Published in India by
Kalpana Shukla
KW Publishers Pvt Ltd
4676/21, First Floor, Ansari Road, Daryaganj, New Delhi 110002
T: +91.11.23263498/43528107 E: knowledgeworld@vsnl.net W: www.kpub.com

ISBN 978-93-81904-13-8

Printed and bound in India

To

My lovely grandmother
(who was my first teacher and probably the best teacher ever),
my endearing parents and
my adorable cousin, Aman,
whose name symbolises 'peace'.

Contents

Acknowledgements

Who says writing is an individual act? There are a gazillion ways one writes; there are a gazillion combinations of factors that mould one's writing, and my writing is no different. Mere presentation of facts and a few arguments and counter-arguments would not have fulfilled the requirement of producing a work that encompasses such a wide array of issues – both controversial and interdisciplinary. The importance of 'my standpoint' and analysis channelled by both theoretical and practical knowledge has been understood and imbibed in writing this book. In this fruitful exercise, a number of individuals and organisations stood by me through thick and thin.

The book would not have seen the light of the day without the endorsement of the man with 'crazy' (in his words) ideas that have contributed most to India's national security architecture. Air Commodore Jasjit Singh, Director-General, Centre for Air Power Studies (CAPS), by beginning the Outreach Fellowship, opened the doors of knowledge, wisdom and innumerable opportunities. Air Marshal A. K. Singh, former AOC-in-C Western Air Command, with his deep-seated interest in the subject helped me pursue the research pathway that I had chosen, and made sure that I never lost track of the central theme.

Dr. Arvind Kumar, Head, Department of Geopolitics and International Relations, Manipal University, put his faith in me when it came to writing this book besides providing me the training to enhance my knowledge in the domain of geopolitics and international relations plus my writing skills. His encouragement enabled me to travel a few times to New Delhi to conduct research, attend conferences as well as present papers on topics that are covered in the book in order to elicit valuable reactions. Professor M. D. Nalapat, Director of Manipal University and UNESCO Peace Chair, has always been an inspirational figure to me due to his exceptional knowledge and intrepid writing, and this book has benefited a lot from all the exchanges I have had with him over the past two years.

My loyalties lie with three incredible personalities who have left no stone unturned to help me reach my destination. From Day One, Group Captain

Manoj Kumar, Senior Fellow, CAPS – a perfectionist, a non-conformist and an impressionist – has witnessed every moment of my journey. He was instrumental in evolving the framework of the book in different ways – his remarkable and prompt peer review of every chapter; his 'diplomatic' way of conveying to me the need to strengthen those parts that seemed too weak to him as a reader as well as to mellow down parts that sounded too strong or even outlandish; the endless conversations with him on the topic even when I was miles away from New Delhi; his unconditional support in terms of exploring new angles, suggesting new ideas and recommending readings, and his unforgettable remarks such as the one on my style of writing – as 'leftist'. I learnt a lot of working and writing with him on issues that are close to our hearts, and as such, I have used this knowledge to the fullest possible extent in this book. I attribute my accomplishment to his perseverance and his undaunted belief in me.

Ms. Cleo Paskal, Associate Fellow, Chatham House and Adjunct Faculty, Department of Geopolitics and International Relations, Manipal University, has had a significant influence on the contents of the book. At a time when I was carried away by the discourse on the climate change negotiations, she opened my eyes by giving me an entirely new perspective on environmental issues that I could imbibe in my work. My numerous interactions with her played a huge factor in sharpening my area of research as well as in providing an 'objective' perspective in the book.

The third person who played a huge role in not only moulding my thoughts on certain relevant issues, such as sustainable development, national security etc., but also in improving my research methods is Dr. Pranav Kumar, Department of Geopolitics and International Relations, Manipal University.

I am extremely grateful to a host of individuals who agreed to take valuable time out of their busy schedules to answer few questions of mine as well as provide germane information and readings. To start with, Mr. Shyam Saran, former Foreign Secretary, former Special Envoy of Prime Minister of India on Climate Change; Mr. Brajesh Mishra, former National Security Advisor, Dr. Anil Kakodkar, former Chairman, Atomic Energy Commission of India; Dr. Kirit Parikh; Dr. Ranjan Chatterjee of the Planning Commission;

Dr. Indrani Chandrasekharan, Adviser (Environment and Forests), Planning Commission; Dr. Prodipto Ghosh, Member, Prime Minister's Council on Climate Change; Dr. Lavanya Rajamani, Professor, Centre for Policy Research (CPR), New Delhi; Dr. Matthew Paterson, Professor of Political Science, University of Ottawa; Dr. Arabinda Mishra of TERI; Dr. Swaran Singh, Professor and Chairperson, Centre for International Politics, Organisation and Disarmament (CIPOD), Jawaharlal Nehru University, New Delhi, and Nitin Sethi, Reporter, The Times of India, New Delhi.

I also thank all the researchers at CAPS who have helped me fine-tune my thought process; in particular Dr. Manpreet Sethi, Senior Fellow, who not only assisted me in the process of formulating an independent view on nuclear energy, but also facilitated my research by providing me several eminent documents, and Air Vice Marshal Kapil Kak, former Additional Director, who supported me in my endeavours and made me realise the significance of my work by calling it my 'magnum opus'. I would like to express my gratitude towards the entire faculty (including Adjunct Visiting Professors) and scholars at the Department of Geopolitics and International Relations, Manipal University who have been highly supportive throughout. I have gained an insight into the Western perspective from brief discussions at a conference held in New Delhi with Rear Admiral Neil Morisetti, Climate And Energy Security Envoy of the UK Ministry of Defence and Foreign and Commonwealth Office; Dennis Taenzler, Head, Climate and Energy Policies, Adelphi, Germany; and Dr. John Jackson Ewing, Research Fellow, Centre for Non-Traditional Security (NTS) Studies, S. Rajaratnam School of International Studies, Nanyang Technological University, Singapore.

I would also like to appreciate the assistance provided by a few environmental (governmental or NGOs) institutions/organisations and libraries that has played a pivotal role in my research. The Department of English, Stella Maris College, taught me the art of writing and critiquing, besides reinforcing my interest in environmental research through the activities of the Enviro Club and the course – Environmental Studies – taken by the passionate and erudite Dr. V. Padma (a.k.a. Mangai). I sincerely thank the Kendriya Vidyalaya Sangathan and all my teachers who gave me the best school education in the country. The industry expertise and extensive support of the KW Publishers gave the final push to transition the project from the final draft to a complete book. Without the proactive cooperation of the library staff at CAPS, TERI, Institute for Defence Studies and Analyses (IDSA), New Delhi, Centre for Science and Environment (CSE), New Delhi, and the Department of Geopolitics and International Relations, Manipal, in providing me the material, I would not have been able to conduct research and produce this book.

Invaluable thanks from the bottom of my heart to all my friends who readily agreed to proofread my work as well as give their comments. Chitralekha Manohar, Manoj Yesodharan, Anand Vidyadharan, Purnima Prakash, Vignesh Ram and Tevita Motulalo made my life easier by taking up this time-consuming task and doing an admirable job. My friends (Chitralekha gets special mention here) also gave me enormous amount of emotional and mental support whenever the going went tough, above all, in overcoming the difficulties I confronted time and again while writing and in keeping myself 'sane' in times of information overload/work overload/ emotional overload/mental overload. My roommate, Rhea Abraham, was instrumental in keeping me healthy by feeding me good food as well as in giving me the space to write by accommodating my erratic schedule.

Most importantly, I sincerely thank my entire family, especially, my parents whose presence in my life makes it worth living. Each achievement in my life is a result of hundreds of efforts made by my parents, and this book is no different.

Preface

The 'green' bandwagon is emerging as one of the biggest profit-making ventures in the world. From energy, to water, to agriculture, to trade, to security, 'green' is—often controversially—permeating the policy. This is not a topic India can ignore. This is not a topic that could be segregated and dealt with exclusively. It is one that needs to be explored and understood, and one that requires leadership. Hence, the title of this book: *Breaking out of the Green House: Indian Leadership in Times of Environmental Change*.

In many cases, understanding the core issues involved requires going back to basics. The term 'environmental security' is self-explanatory in many ways, but the core issues that drive the discourse on it can be understood only if one goes deeper into the basics of theory and practice of the subject. For example, environmental change, which is a key aspect of environmental security, is considered a non-traditional security issue; however, environmental crises such as famines, floods, and fires are some of the most traditional security challenges societies face. From the time, millennia ago, when the early Indus Valley civilisations grew up around control of their environment (in their case through advanced irrigation techniques), the link between our security, and prosperity, and our environment has been clear.

Additionally, understanding the 'security' implications, traditional or otherwise, of environmental change requires disentangling the distinction between 'human' security and 'national' security. A critique of 'nation' would debunk the state construct of national security that has given the state the legitimacy to spread its tentacles over the roots of human existence. A critique of 'human' would subject human existence to the existence of the state—the overarching structure that sustains modern human existence. At the end of the day, this debate could be easily put to rest by drawing parallels between the 'self-vs-other' debate; it essentially leads to the understanding that self is a reflection of the other, and the other is a reflection of the self. In short, they coexist. Similarly, national security and human security cannot exist in isolation. At a time when security has assumed multiple dimensions including cultural and psychological, it is naive to distinguish between traditional and non-traditional security threats and prioritise them in such a manner that the latter category of threats ends up receiving second class treatment.

Environment has also permeated the diplomatic and the strategic circles as the international community expends more and more time and energy on 'achieving consensuses' and 'drafting treaties'. Whether or not one wants to securitise the most 'scared' thing on earth, which is life, one simply cannot run away from the threat of environmental change that both states and individuals have already begun to face.

Another core question is, has the green house gas theory failed to explain the phenomenon of environmental security/insecurity? For the purposes of this book, the answer is: do not blame the theory; blame the practitioners. The 'green house' is a constraining construct that has to be broken by the practitioners in order to focus on more complex environmental change issues by going beyond the discourse stipulated by the state. The question of whether it is a natural construct or a human construct is just one strand of the story. The question of 'now' or 'later' in terms of action is yet another crucial element that is being exploited astutely in order to delay the act of doing a world of good for the future. Therefore, the title of this book, breaking out of the green house, implies two things—first, to break the traditional stereotypes of geopolitical thinking, and second, to act on environmental issues. Since the book covers only the state-driven debate, it is one-sided to a large extent, albeit it opens the doors for further research in the field of environment from a perspective that questions the state's stake in it. The Indian leadership forms the fulcrum of the book's argument.

Of course, there is a yawning gap between 'what is' and 'what should be' but as long as the probability of attaining 'what should be' is bright, we should continue to brainstorm ways to reach that point. It may not be utopia, but it might help in making the world 'secure'—the ultimate dream of the political, scientific and strategic communities. We should strive to infuse some credibility into the most abused phrase in international relations—'maintenance of international peace and security'. This is the reason I have dedicated all my chapters to our policies and strategies (such as they are) on environmental issues, including their drivers at all levels—domestic/national, regional and international.

The environment is changing but the pace and manner in which it is changing, is difficult to ascertain. India is one country that has the capacity for spawning a zillion solutions for the zillion problems faced by the world currently, including environmental change. What is needed is sheer, or even mere, political will.

On a very personal note, the past has taken me one step closer to the future with the present complementing the events of both. When I set out to understand environmental issues during my college days, everything seemed to be open-and-shut cases because first, environmental degradation was undoubtedly a reality, and second, the question as to who would take something as critical as environmental degradation lightly.

However, the more I came to understand the overlapping complexities and interests involved, the more I realised that true understanding, which could lead to lasting security, would have to encompass a much wider range of issues. It is also worth noting that throughout this book, I have assumed that the environment does not have a life of its own, and humans need to 'make' the environment conducive for survival and sustenance. I'm sure Machiavelli and Hobbes would second my 'sweeping statement' that humans are selfish.

From an object of curiosity, to a subject of interest, to a part of my life, the environment and its web of interrelated issues, ranging from economics to politics to sheer survival, has become more than just passion for me. The journey of writing this book rekindled some old memories of my association with the environment; it brought me closer to the environment, and helped me be *au fait* with the politics that has managed to hijack this sacrosanct 'being' that sustains everything on earth. I also realised that language was so constrained, and at times deceiving, that I could hardly express the thoughts that kept floating into my mind.

What should follow this first step is a deeper understanding of the issues of environmental politics, security, diplomacy, strategy and sustainability to set a bottom-up, trans-national agenda in motion. A field that has so many stakeholders requires special attention particularly since it is futuristic (thus, strategic), and the development of an Indian perspective on these issues is critical not only for our growing ambitions in the international system, but for our very security.

Environmental Change and it's Global Implications: An Introductory Essay

Jasjit Singh

"We do not inherit the earth from our ancestors,
We borrow it from our children."

— Red Indian Proverb

The story of Noah's Ark many millennia ago, in the hoary past of human civilisation, is not easy to explain in the context of the modern world. There can be little doubt that if it would have taken place, there would have unquestionably been a major environmental disaster that would have triggered the circumstances where only the minimum living humans and animals could have been saved to let the world grow into its future. We have more certain empirical historical evidence of many civilisations—the Akkatians, the Romans, the Indus Valley, the Mayans—collapsing because of environmental change. It can be accurately argued that political conflicts, political rivalries, and economic discord contributed in no small measures to the process of the decline of these civilisations. There is sufficient historical and empirical evidence to prove that environmental change had probably been the dominant factor in the decline and demise of the Mayan civilisation.

The historical accounts and numerous scientific studies, especially on climate change due to global warming, were literally dismissed in the 1980s. If at all, they were given limited attention because the environmental threat was not perceived to be imminent or catastrophic; then was a world

---

Air Commodore **Jasjit Singh** AVSM VrC VM, IAF (Retd), awarded the Padma Bhushan for a life-time's contribution to national security and defence, is the Director General, Centre for Air Power Studies, New Delhi.

where human occupational focus was to enhance economic growth, consumerism, and very high levels of resource consumption to achieve the goals of being or trying to reach the levels of advanced industrial countries. As it is, the Industrial Revolution had become the instrument of an imperial age and global colonisation that brought with it immense challenges, the largest perhaps being what Kennedy called the 'Imperial Overstretch'. And colonialism, which consumed large amount of resources from the colonies and high consumption in the colonial countries for their development and higher 'quality of life' over nearly three centuries, proved that the native American proverb (cited above) was simply ignored risking the future to gain advantages in the present.

The predatory 'advanced and developed countries' have tasted success in their agenda of constructing an apparent paradise. Now, the basic principle of human life—the continuation of human life on earth itself—is now facing the prospect of the unpredictable vagaries of environmental change and degradation, which have begun to manifest themselves in many ways at a greater pace than what was even a decade ago. At the minimum, there is a rapid rise in natural disasters—changing patterns of climate change, droughts, earthquake/tsunamis, cyclones, and so on. Melting of the polar ice and numerous glaciers is seen accepted as a major threat. On the other hand, climate change has so far proved to be beyond the human capacity to control science, which the human race has been trying to master since ages. The scientific communities' findings in terms of increasing carbon dioxide, rising sea levels, accelerating melting of glaciers have enhanced the perceptions and knowledge of the international community that the disasters may be much greater and closer in time than earlier believed. But given the five century-old belief in sovereignty and the organisation of sovereign nation states pursuing their national interests, and the uneven development of nations across the world (and hence their consumption patterns), the problem of reaching agreements leave alone consensus and implementing solutions is bound to be a long drawn one.

On the other hand, international organisations like the Inter-governmental Panel on Climate Change (IPCC), NASA, and ISRO (Indian Space Research Organisation) have been keeping an eye on the developments in the climate change domain; these organisations started to receive the attention of the international community. In fact, ISRO has found rather alarming impact of climate change on the Himalayan glaciers, agricultural yield, and hydrology in India. Given the known trends, the strategic implications of environmental change are phenomenal for all countries, where the developing countries like India would face more challenges that are serious since it requires greater energy and resource consumption, has limited capabilities for developing new technologies; yet, India must increase consumption in order to improve the quality of life of 1.2 billion people. In fact, the interests of China, India, Bangladesh, and the US in the Indian Ocean region (which provide the source of the monsoon not to talk of other factors) could be affected adversely with the rise of sea-level and storm surges. These, in turn, will impact all the major and minor naval assets and commercial ports, including the low lying land areas like the Maldives. With Diego Garcia just a few feet above sea level like Maldives, the US has to adapt this base to the impact of environmental change.

India, in fact, has to take into consideration a multitude of factors while framing its local and regional policies and actions related to climate change while at the same time, safeguard and promote its own interests among the international community. Because of the fact that the country was de-industrialised during the three decades during colonisation, it is still far down the scale in development processes; India faces fundamental weaknesses in dealing with required policies. Inevitably, it faces a serious challenge of prioritisation between faster techno-economic and social development, which demand even higher energy consumption, and the imperative of managing climate change. The challenge is made more serious because contrary to conventional wisdom, it has a major deficit in terms of natural resources. Of course, it has one of the largest reserves of coal in the world; but the coal has very high ash content. Therefore, the options are to import coal for its thermal power stations (and other uses) and/or employ technologies for cleaner coal; these technologies are expensive and/or not easily available and affordable. It is not surprising; therefore, more than 45 per cent of coal used in power stations is imported from Australia and Indonesia. Similarly, other hydrocarbon consumptions has gone up and we are now importing around 72 per cent of our requirement of crude oil, which is expected to go up as much as 85 per cent or so in the next decade.

It is this imperative need for more energy to ensure high economic and industrial/agricultural growth on one side, and the implications of increasing need for hydrocarbons on the other has led to seek greater availability of environmental-friendly nuclear energy. It could only be done with import of uranium fuel for power reactors since the total reserves of uranium are unlikely to support more than 10,000 MW capacity. Hence, the importance of Indo-US agreement of 2008 will go a long way in harmonising environmental policies with development priorities. India will probably be self-reliant on nuclear power when the Fast Breeder reactors come on stream, which would solve the problem of nuclear waste and its negative impact on environment. The experimental Fast Breeder Reactor has been working satisfactorily for two decades, and the prototype reactor will soon go on stream. However, to get significant reactor capacity with Fast Breeder Systems, it may take another three odd decades. Then we can start thinking of reducing the use of hydrocarbons at least for energy. Again, the next three decades are also crucial for India's development at a high rate if it has to seriously avoid lagging behind not only developed countries but also China.

Meanwhile, India has to open floodgates leading to green technology (which can be affordable only with high economic growth) in every sector including power and transportation as the author has stressed in this book. Alan M. Eddison of Green Earth Affairs had said, "Modern technology owes ecology an apology," to argue the same point. Nevertheless, India has a challenge to match economic growth with growth of green technology. India still has a long way to go in eradicating poverty on one side, and pursue sustainable development on the other. For example, the risk assessment of infrastructure building in Mumbai, the financial capital of India just above the sea level (like China has Shanghai), reveals that the places like Mumbai face sinking problems due to population growth and modernisation. Hence, the adoption of all-inclusive development policies that could augment India's adaptive capacity and assist its mitigation activities; these require enduring political will and harmonisation of various departments of the government that is also adjusted to our federal political structure, which at time has begun to be a handicap.

The battles about environmental security have to be largely fought at the international level through diplomacy. Climate change had become the centre of major discourse by the 1980s, and negotiations have now become the fulcrum of debate surrounding environmental and development issues justifiably as well as unjustifiably in many cases; these battles are no less complex or fought with lesser skill and courage than those with weapons on the battlefields. In both the cases, knowledge domain assumes a critical role. After the creation of the United Nations Framework Convention on Climate Change (UNFCC), the Kyoto Protocol has probably been the most and only noteworthy milestone in climate change negotiations. The political fault-lines between the developed and the developing countries, and within the developing countries—because of their higher differential state of development—between the energy deficient and energy exporting countries, and between politicians and scientists—have often derailed negotiations from time to time so much so that the Kyoto Protocol failed to deliver the results that were initially envisaged.

Environmental change is not restricted to climate change; the latter is only a one crucial component of the former. Its importance may be judged from the fact that it affects and concerns the world, and the international community has to accord it the highest priority for the sake of the future of the human race and civilisation. Time has also come to pay attention to non-climate change related issues, such as urbanisation (which is increasing rapidly in India and most other countries leaving serious deficit in infrastructure), deforestation, ground water depletion, pollution, stress in the availability of resources, and so on as each of them degrade the environment and impact security of the present and future generations. India and its neighbouring countries face serious problems related to environmental issues, such as loss of territory, lack of fresh water in the Maldives, drying up of River Indus, and China's policies on river waters as a riparian state, which could adversely affect India and Bangladesh. While it is in India's interest to pay serious attention to environmental change and its adverse effects, it also would be well advised to bring in the neighbours into a regional policy although the neighbouring countries may not like India leading such initiatives. This aspect perhaps could be left to country like the Maldives, or each country could take on responsibility of some aspects on a regional basis with India's support, and bringing in the UN agencies wherever possible and relevant.

This volume is the result of extensive objective research and study by a post-graduate young scholar at the Centre for Air Power Study, Ms. Dhanasree Jayaram. She acquired her Masters degree from Manipal University, and is a refreshing example of a young scholar producing an exceptionally brilliant volume in an area—complex and multi-disciplinary—many older people would avoid getting into. She also proves that grey hair and grey matter are not necessarily linked to each other, as many of us in this country believe as conventional wisdom! I feel privileged to introduce this excellent study to the readers.

The foregoing and many other issues of concern regarding environmental changes are matters of India's long-term security, and the challenge is great especially since it has to be harmonised with other policies impacting eradication of poverty, increasing the economic growth rate, greater access to energy, which would provide sustainable development, and so on. We hope that the reader—be he or she a policy maker, opinion shaper, a scholar, a military professional, or a layman concerned bout the country's future as well his family's future generations and their welfare—would enjoy this highly readable and future policy-oriented volume.

1

The Changing Patterns of Global Climate Change Architecture

> "It is the responsibility of each human being today to choose between the force of darkness and the force of light. We must, therefore, transform our attitude, and adopt a renewed respect for the superior laws of the DIVINE NATURE."

These words of Maurice Strong, the former Secretary General of the UN Environment Programme at the 1992 Earth Summit sparked off a fresh debate that forced the international polity out of slumber. If the 20th century was marred by the World Wars and other conflicts that slaughtered millions of lives, the 21st century is turning out to be the century of environmental wrath that could take even more lives than any form of conflict. The human-induced process of environmental change that was set off way back during the era of civilisation-building took a perilous turn in the 20th century due to the industrialisation drive of the Western countries; it is now gradually worsening as the international community awaits a consensual deal. Climate change is one such threat, and among all environmental threats, this is one phenomenon which has gathered the most attention. One could say that it is not the definition of climate change that needs to be looked into; it is the definition of the 'response to climate change' that needs to be transformed so that the 'measures' bear fruit. Geopolitically, climate change is an issue that encompasses all facets of international politics that steer a nation's march towards achieving national comprehensive power. Ironically, the international negotiations themselves have become synonymous with 'conflict' rather than 'cooperation'. Resolution of the geopolitical chaos is the key to the effective tackling of the effects of climate change. To be more futuristic, and perhaps idealistic, the negation of the geopolitical chaos and the introduction of an alternative such as technology cutting across political fault-lines should be given due attention. The first step towards that would be re-standardisation of international relations that would create political willingness.

India is one of the most vulnerable countries in the world not only due to its geographical conditions but also due to its weak infrastructure. At the international level, as an important participant on the negotiating table, India has come a long way from being a silent spectator to being a leading voice from the developing world. However, the road ahead seems challenging as in an increasingly competitive world, with wealth comes power, with power comes responsibility, and with responsibility comes liability. At the national and local levels, it is imperative to examine the methodology of adopting a robust climate change policy that simultaneously deals with three issues—energy management, development and reduction in emissions of pollutants, and green house gases—since the country is grappling with poverty and energy deficit. The effort in the succeeding chapters would be to dissect the poles that exist in this country apart from giving a perspective on the futuristic scenario in terms of sustainable development, adaptation, mitigation as well as energy security. These poles are being dictated by the differing stances of the political class, the judiciary and activists. As part of an India-centric evaluation of the impact of climate change, the chapter would also delve into the intricacies of national security that could be threatened in various ways. As the world gears up to deal with the vagaries of climate change —the crux of environment change – both the theory and practice of the politics of climate change would be revisited in the subsequent pages. This could map out the steps that could tighten up global efforts in the areas of adaptation and mitigation. These physical and metaphysical issues are hard to disentangle like the Hegelian dialectic but an effort would be made to reach the 'syntheses' through analysis and case studies. This chapter would mostly restrict itself to the existing narratives on climate change, which is Western-oriented in order to build the foundation for the Indian perspective on the subject. The first and foremost task is to understand climate change as it is, and how it is perceived through the prism of international relations.

Redefining Climate Change

A Multi-faceted Phenomenon

Climate change is a dynamic phenomenon that has multiple layers. To classify it as merely scientific or meteorological will be too simplistic. In today's world, every issue develops tentacles that reach out to various spheres, usually guided by conflicting ideologies. This is the reason why definitions have become transient and relative. Though climate change is incontrovertibly a global phenomenon that has universal causes and consequences proven by the scientific community time and again, it affects different entities or spheres making the task of framing the exact definition highly contentious. In geopolitics, this becomes even more complicated since the world is clearly divided by nation states with lucidly delineated boundaries and cogently defined national interests.

On the one hand, the international community represented by various organisations have come up with definitions that have got worldwide acceptance. But one has to take into account the politics of climate change which guides the fundamental theories related to it; the former has assumed significance to such an extent that the real dangers of climate change are being overlooked. The starting point of the differences between the perspectives of the scientific and the political communities is the definition itself exemplified by the Inter-governmental Panel on Climate Change (IPCC) and the United Nations Framework Convention on Climate Change (UNFCCC) respectively. The latter defines 'climate change' as "A change of climate which is attributed directly or indirectly to human activity that alters the composition of the global atmosphere and which is in addition to natural climate variability observed over comparable time periods."[1] The former states, "Climate change refers to a statistically significant variation in either the mean state of the climate or in its variability, persisting for an extended period (typically decades or longer). Climate change may be due to natural internal processes or external forcings, or to persistent anthropogenic changes in the composition of the atmosphere or in land use."[2] While the UNFCCC specifically focuses on climate change attributable to human activities for obvious reasons, the IPCC looks at it more holistically by making a distinction between 'climate change' that is triggered by human activities and 'climate variability' that is ascribed to natural causes. It is the human activity that has divided the world into several blocs, and the definitions yet again vary from bloc to bloc since the degree of impact and the urgency of response to climate change are variable.

Geopolitics and International Relations Approach

Geopoliticians draw conclusions based on these diverse perspectives of climate change, and thus, expand the definition to include the facet of international security. According to them, it is the architect of the process of redrawing the world map besides being the platform for the 'clash of interests' between nation states. In international relations and world politics, environment has always been one of the top priorities, and with the advent of the concept of climate change, it has assumed even more significance. The blame-game over the origin of climate change and the debate over the possible measures to tackle it have turned it into a playground of power politics. Just as it has divided nation states, it has also divided the international theorists. The debate between the 'modernists' and the 'ecoradicals' is the most noteworthy one. The former group is of the belief that environmental protection cannot be achieved at the expense of development. Thus, they champion the cause of environmental protection and promotion of the development agenda concurrently. This model essentially lays emphasis on the upgradation of the existing scientific and technological resources that could mitigate environmental degradation. However, these environment-friendly techniques have been rejected by the 'ecoradicals' according to whom the only way to decelerate the dramatic rise in pollution and exhaustion of natural resources is to halt the process of growth itself. This was also the central finding of a group of scholars of Club of Rome who published 'Limits to Growth' in 1972. They strongly believe that any progress in science and technology cannot solve the problems that the humanity will face since the ecosystem does not have the capacity to stand 'human atrocities' beyond a certain point. They call for a complete overhaul of human nature, lifestyle and population control measures.[3] These two theories are the two sides of the same coin. The first one accepts the reality of the current economic growth model that is central to any country's comprehensive power theory but fails to make an assessment of the notion of development. Though it promises to provide long-term solutions, it fails to gauge the status of the environment in the long run. The second one takes a more idealistic stand that regards the whole world as one organism, and the destruction of one part would result in the destruction of the entire planet. It completely excludes the needs of a growing and demanding population. Though it has a long-term perspective, it fails to give an alternative to the current economic model apart from demanding decentralisation, devolution and diffusion of the powers of the nation state in order to generate 'ecocentric' communities. New environmentalism is deemed an offshoot of the emerging threats of global warming and climate change. The theory of new environmentalism recognises the fact that global warming is a symptom and not a problem; the problem lies in the wasteful and exploitative nature of human society.

Science Matters

A clear understanding of the theories associated with climate change should be backed by scientific and experiential facts. It is a well-known fact by now that the root cause of climate change is the 'greenhouse effect'. This is the primary factor that contributes to the sustenance of life on earth as it keeps the temperature of the earth sufficiently high. But the same effect could prove disastrous if it is intensified by carbon emissions due to burning of fossil fuels and deforestation. Methane emissions caused by agricultural activities are also said to be significant contributors although this has become a major bone of contention between the developed and developing countries. The primary greenhouse gases include carbon dioxide (CO_2), methane (CH_4), nitrous oxide and halocarbons, out of which CO_2 is the biggest contributor. Human activities, as a source of climate change, have been highlighted by the IPCC in their Assessment Reports, and in their Fourth Assessment Report (2007) they reiterated that 'global GHG emissions due to human activities have grown since pre-industrial times, with an increase of 70 per cent between 1970 and 2004'. They went one step ahead to increase the probability to more than 90 per cent that reinforces the chief cause of the increased greenhouse effect to be human activities.[4] The natural process by which the carbon is released into the atmosphere is accelerated by the activities of human beings. This causes global warming, which in turn sets off radical variations in the weather including precipitation, winds, temperature, and so on. For example, an increased concentration of CO_2 in the atmosphere is known to affect cloud cover. Over the years, the sceptics have discarded the findings of these scientists sporadically and come up with their own conclusions that consider climate change a natural phenomenon or attribute it to other sources such as cosmic rays that affect the formation of clouds. A case in point is the reduction in solar activity, which is regarded by a majority of the climate scientists as the cause for the Little Ice Age between 1650 and 1850. One of the NASA studies has refuted this claim by stressing on the fact that the average amount of energy emitted by the sun has neither decreased nor increased drastically since 1750, and that only the lower parts of the atmosphere have become warmer while the upper atmosphere has remained cool as the GHGs trap heat in the former.[5] Since the politics of climate change revolves around the assumption that human activities are the dominant contributor, climate scepticism loses relevance in this context to a certain extent.

Impact of Climate Change: An Indian Perspective

Facts and Figures as Seen Through the Indian Eye

Climate change could have unprecedented environmental, social and economic consequences, which in turn hold the potential to create a shift in the relations between different nation states and eventually culminate in an environmental or political semi-apocalypse. Bill Clinton very famously said, "First, I worry about climate change. It's the only thing that I believe has the power to fundamentally end the march of civilisation as we know it, and make a lot of the other efforts that we're making irrelevant and impossible."[6] The real signs of a changing climate are manifested by sea level rise, global temperature rise, shrinking ice sheets, glacial retreat, extreme weather events, and so on. Some of the figures released by NASA are grim reminders of what is in store for humankind if immediate measures are not taken to mitigate climate change. For instance, the CO_2 concentrations are at their highest in 6,50,000 years. The global average sea level has risen by 4-8 inches over the past century, and it is increasing at the rate of 3.27 mm per year. There has been an average increase of 1.5°F in the global temperature since 1880. Besides the doubling of the loss of Greenland Ice between 1996 and 2005, January 2000 to December 2009 was recorded as the warmest decade.[7] Indian scientists have also added more credence to these findings as evidenced by the reports of the ISRO that concluded the retreat of the Himalayan glaciers by nearly 16 per cent in the last five decades. Their study on the impact of climate change on agricultural yield and hydrology also revealed disturbing results. In the case of the former, it is in the form of reduction in the output of wheat, rice, maize and pearl millet while in the case of the latter, it is in the form of increase of runoff in almost all the major river basins of India in the month of June. The 2009-10 report of the ISRO also said that Indian livestock are a major source of CH_4 emissions.[8] A rise in temperature in the Tibetan plateau could reshape the sub-continental river systems especially that of the Brahmaputra and the Indus, and convert them into monsoon-fed rivers that could annihilate a significant proportion of the population in South Asia including India.

Indications: India and the World

Scientific evidence shows that climate change leads to higher frequency of extreme weather events such as heat waves, heavy precipitation events, floods, droughts, fires, and severe cyclonic storms, which in turn result in a surge in deaths, injuries, toxic contamination, infectious diseases, social disruption, environmentally forced migration and the list goes on. A simple illustration would be the implications of higher temperatures; they could heighten both flood and drought risks, the former due to a sustained increase in precipitation and the latter due to an increase in evaporation. There are countless indications which cannot be thwarted by the sceptics. These are the unusual heat waves in North America, Russia and Europe; the unpredictable precipitation episodes in Latin America and South Asia; hurricanes in the US; cyclones in the Indian Ocean region; droughts and desertification in Africa, West Asia and North Asia. As far as India is concerned, the Global Climate Risk Index 2010 released in Copenhagen revealed that the country has witnessed 325 extreme weather events in the last 18 years in which

3,255 people have died annually.[9] One could easily conclude that although sceptics have raised doubts over the cause of intense and more frequent El Nino and La Nina events, they could easily be considered the signs of a larger catastrophe that climate change could unleash in future. Moreover, the IPCC states, "Taken as a whole, the range of published evidence indicates that the net damage costs of climate change are likely to be significant and to increase over time."[10]

The rise in global sea levels caused by the melting of polar ice caps and glaciers is expected to result in the submergence of low-lying areas including river deltas, coastlines and small islands. For example, the IPCC has predicted that if the sea level rise exceeds 2 m, nearly 50 per cent of Maldives would be inundated and in the case of Bangladesh, a 1.5 m rise would submerge 15 per cent of all land area and about 20 per cent of farmland, with 20 per cent less in both agricultural production and the nation's total population.[11] It is expected to have adverse impact along 7,000 sq km of the Indian coastline. A 1 m rise in sea-level would lead to the displacement of nearly 40 million people in India.[12] Interestingly, the rising sea water has ended the dispute between India and Bangladesh over the New Moore Island by engulfing it in 2010. Massive displacements were caused by the sinking of two other islands in the Sunderbans—Lohachara and Ghoramara. The changing patterns of salinity caused by climate change has affected not only the mangroves but also fishing, leading to the disappearance of one of the mangrove species called Sundari trees, and economic enslavement of the fisherfolk. Besides, an increase in the global temperature could affect the ecosystems, which would disrupt habitats such as coral reefs and alpine meadows, which in turn, would lead to the extinction of several plant and animal species. The dwindling number of Royal Bengal tigers has been attributed to rising sea levels as their migration to higher ground would be prevented by human habitation in the adjoining lands. The effects of climate change on the society and the economy are, therefore, expected to be severe. If one takes into account 'the cost to agriculture, forestry, water supply, fisheries, infrastructure, energy, hurricane damage, drought damage, land loss, loss of wetlands, loss of forest, loss of human life, loss of species as well as pollution and migration, the total annual cost of all the aforesaid global warming effects is estimated to average 1.5-2 per cent of the current global GDP, that is, between US$ 480 and 640 billion'.[13]

Climate Change and National Security: The Indian Viewpoint as Painted by an Outsider

The inter-connectedness between climate change and national security has been a source of debate for policy-makers across the globe. It is preposterous to dissociate national security from human security. If the population of a nation state is not socially, economically, politically and psychologically secure, the national security of that particular nation state would remain incomplete. National security encompasses not only military strength, but also social security, environmental security, energy security, economic security and a wide range of aspects that impinge on human security in general. After all, the nation state is just a conglomerate of its inhabitants at the atomic level; yet, policy-makers have been trying to downplay the threat that climate change poses to national security especially in a country like India. This is the reason why attempts have been made by the research community to link the two so that the issue of climate change is taken as seriously as the menace of terrorism. In India's context, there are various issues that imperil our national security but very often they are straight-jacketed into our neighbours on the west and the north.

Besides, the pulls and pushes in the country that emanate from different entities including the Government, the industry, the judiciary and environmentalists have also led to myriad contradictions at the policy-making level. The development versus environment debate has ensured that India's policies towards achieving a low-carbon economy and sustainable development remain in a limbo. The international pressure to adopt binding emission cuts on the one hand and the domestic requirement of energy, on the other, have led to sharp divisions within the country. Securitisation of the debate is more or less a taboo as evinced by one of the premier environment journalists in the country, according to whom 'playing the waiting game' is going to be the strategy of the majority of nation states including India. Similar indications have been made by the country's policy-makers too according to whom the connection between climate change and national security is not direct, and the question of sacrificing the present for the future is unthinkable as far as sustainable development is concerned. The international community has so far not shown the keenness to go beyond the ambit of 'short-term solutions' and 'low-hanging fruits'.

Climate change has the 'potential' to add a whole new dimension to India's disputes with its neighbours. The first and foremost issue that comes to the mind of every Indian in this context is the Kashmir dispute. The world's highest battlefield, the Siachen glacier, is said to have receded by nearly 800 m in the past 20 years due to climate change. It has been found that the military presence in the region has resulted not only in the deterioration of the ecosystems but also in the pollution of the Indus River.[14] Interestingly, Pervez Musharraf, the former President of Pakistan claimed in his dissertation that the Kashmir dispute was primarily based on the distribution of the Indus waters between India and Pakistan. Stressing on the fair distribution of waters, he asserts, "If one were resolved, the other would not exist."[15] It is highly possible that if the Indus River dries up due to global warming, a water-deficient Pakistan might resort to a full-fledged war with India over the distribution of its waters. Pakistan has already made it clear that it would not hesitate to use its nuclear weapons against India if the latter chokes the water supply to its territory as confirmed by a report prepared by nuclear physicists Paolo Cotta-Ramusino and Maurizio Martellini of Landau Network, Italy, on Pakistan's nuclear policy in which they quote Lieutenant-General (Retd) Khalid Kidwai, the Director-General of the Strategic Plans Division of Pakistan.[16]

The threats of water wars or a Third World War due to water scarcity are palpable for not only India and Pakistan, but for the entire world. Water wars in the Middle East/West Asia due to scarcity of water, particularly between Israel and Palestine, are quite well-known. If the trends of drought, famine and disease continue, the chances of population shifts and political turmoil could increase further in this region. Since economic, energy and climate challenges are closely inter-linked, the security implications are graver. Countries with fragile governments could become failed states. Some of the countries in Africa are deeply vulnerable to this threat since they are already plagued by internal conflicts such as civil wars. For example, a

18-month study of Sudan by the UN Environment Programme reveals that depreciation in the amount of rainfall, decrease in crop yields and increased desertification have resulted in the exacerbation of the conflict between north and south Sudan.[17] The conflict has culminated in a permanent split between the two with the separation of southern Sudan based on a referendum in 2011. Climate change could take away the livelihoods of millions of people, especially those dependent on agriculture and fisheries, who might find an alternative in anti-social activities. Another indication of the geopolitical consequences of climate change is the ongoing tussle over the Arctic region. The shrinking of the ice cap has resulted in the possibility of exploration of natural resources as well as opening up of new shipping lanes in the region, which could give rise to conflicting claims and tensions with regard to maritime territories and exclusive economic zones as reiterated by the United Nations Security Council statement released on July 20, 2011.[18] Countries including Canada, Russia, the US and Denmark have already started to lay claims in the region. The whole concept of maritime boundaries will have to be redefined in case of submergence of coastal areas and islands, especially if they are of strategic significance.

Coming to India, migration is the single biggest threat to its national security. Rising sea levels and glacial recession is already leading to inter-regional migration. If the effects of climate change intensify, the coastal areas of Sri Lanka in the North and the East that are Tamil dominated, could be affected, leading to their migration to India. Although the civil war has come to an end in Nepal, the glacial recession and extreme weather patterns in Nepal could also result in mass migrations to India. India might have to make its transit laws stringent in order to curb an influx of migrants. However, migrations from the Maldives and Bangladesh would be the biggest challenges for the countries in the region. The north-eastern states and some of the metropolitan cities like Mumbai are already flooded with Bangladeshi immigrants owing to different reasons—cyclones, frequent flooding and the civil war of 1971—among others. The disappearance of the Maldives could force its people to find shelter in countries across the world, especially India. However, the linkage between climate change and migration has not been appreciated by many analysts as according to them, migration caused by poverty and unemployment is induced or exacerbated by climate change.

Strategic Implications

One could also evaluate the impact of climate change on the strategic interests of the major powers to interlink climate change with national interest. The most appropriate case study would be the Indian Ocean strategies of the US, India and China. The US has its largest base at Diego Garcia (Camp Justice) – an Indian Ocean island – which is only a few feet above sea level. Sea level rise could lead to the submergence of this island, taking away a critical US military staging area. Moreover, the US believes that climate change could lead to conflicts that would entail more American involvement. A recent report by the American Security Project, an advisory group of high-powered Republicans and Democrats, called global warming 'not simply about saving polar bears or preserving beautiful mountain glaciers ... (but) a threat to our security'.[19] As far as India is concerned, all its naval bases situated along the Indian Ocean rim—Cochin, Mumbai, Karwar, Visakhapatnam and others – would be afflicted marginally by the rise in sea level.

However, the most affected could be the Joint Forces Command at Andaman and Nicobar Islands and the Car Nicobar Air Force Base since the islands are also susceptible to tropical cyclones and tsunamis (though the former is considered safer). Thus, India's coastal security could be jeopardised even in terms of terrorist attacks, and the overall defence structure of the country could be adversely affected, further aggravating the national security. The security implications of climate change will also hamper China's growing assertiveness in the region. Its deep water ports and refuelling stations in Gwadar, Hambantota, Chittagong, Sittwe and Kyaukphyu are situated along the most vulnerable coastlines. However, these conjectures do not have the potential to alter the overall geopolitical rivalries between these nations. At the same time, the efforts to protect these coastal installations through various ways including raising them even further above the current sea-level, so that they are not affected by the sea-level rise, would prove to be expensive and short-term solutions.

Loose Ends Tied Together

The divisions and sub-divisions, causes and consequences, perceptions and misperceptions, solutions and resolutions are inherent in the global climate change debate. For the emerging countries, climate change is an economic issue of developmental concerns; for industrialised countries, it has more or less become a political-cum-ethical issue concerning raison d'état and developmental supremacy; and for the least developed countries, it is a human rights issue that endangers their survival; this will be dealt with in detail in the third chapter. For the theorists—modernists and ecoradicals—it is a solvable issue through distinct approaches but they have been largely inadequate in terms of practicality. Climate change, whether caused by human activities or natural processes, is an issue of concern, and thus, needs to be one of the top priorities for the policy-makers. Climate scepticism is a regressive movement that would only lead to the false belief that exploitation of the environment is the birthright of human beings. There are innumerable indications and studies that should clear the doubts in the minds of sceptics. India is definitely one of the biggest victims of these signals, and the future does not look bright. Not only does climate change affect its socio-economic sectors, it clearly has an adverse impact on its national security that is closely linked to the society and the economy as well as its strategic interests.

World leaders across the globe are trying different means to articulate their stands on climate change; for example, a cabinet meeting held under water by Maldives in 2009, a meeting held by Nepal at the altitude of 17,192 ft at the base of Mt. Everest, and a cabinet meeting held by the Mongolian ministers in the middle of the Gobi desert, wearing 'Save the Planet' baseball hats in 2010. India is also doing its bit to not only galvanise the international community towards a collective arrangement but also to carry out programmes within the confines of its boundaries. But is it enough?

Notes

1. The United Nations Framework Convention on Climate Change, Article 1 (For the purposes of this Convention), Paragraph 2, see http://unfccc.int/essential_background/convention/background/items/2536.php, accessed on March 18, 2011.

2. IPCC Working Group I: The Scientific Basis, Appendix I – Glossary, see http://www.ipcc.ch/ipccreports/tar/wg1/518.htm, accessed on March 18, 2011.
3. Robert H. Jackson and Georg Sørensen, *Introduction to International Relations: Theories and Approaches* (UK: Oxford University Press, 2007), pp. 261-263.
4. IPCC Fourth Assessment Report, Summary for Policymakers, in *Climate Change 2007: Synthesis Report*, approved in detail at IPCC Plenary XXVII (Spain: Valencia, November 12-17, 2007), p. 5, see http://www.ipcc.ch/pdf/assessment-report/ar4/syr/ar4_syr_spm.pdf, accessed on June 13, 2011.
5. "Causes: A Blanket around the Sun," *Global Climate Change: Vital Signs of the Planet*, National Aeronautics and Space Administration, see http://climate.nasa.gov/causes/, accessed on June 12, 2011.
6. See, Dan Perry, "Clinton: Climate Change is the World's Biggest Worry," http://www.newsday.com/news/local/wire/newyork/ny-bc-ny--worldforum-clinto0128jan28,0,379601.story?coll=ny-region-apnewyork, accessed on February 24, 2011.
7. "Global Climate Change: Vital Signs of the Planet," National Aeronautics and Space Administration, see http://climate.nasa.gov/, accessed on June 12, 2011.
8. "Himalayan Glaciers Retreated by 16% in Nearly Five Decades: ISRO," *Daily News & Analysis* (Bengaluru: March 27, 2010), see http://www.dnaindia.com/india/report_himalayan-glaciers-retreated-by-16pct-in-nearly-5-decades-isro_1364163, accessed on June 16, 2011.
9. Sven Harmeling, "Global Climate Risk Index 2010: Who is Most Vulnerable? Weather-Related Loss Events Since 1990 And How Copenhagen Needs to Respond," *Germanwatch* (Berlin: December 2009), p. 11, see http://www.germanwatch.org/klima/cri2010.pdf, accessed on June 16, 2011.
10. IPCC Fourth Assessment Report, Summary for Policymakers, in *Climate Change 2007: Impacts, Adaptation and Vulnerability. Contribution of Working Group II to the Fourth Assessment Report of the Intergovernmental Panel on Climate Change* (UK: Oxford University Press, 2007), p. 17, see http://www.ipcc.ch/publications_and_data/publications_ipcc_fourth_assessment_report_wg2_report_impacts_adaptation_and_vulnerability.htm, accessed on June 15, 2011.
11. R. K. Gupta, "Global Warming and its Effects," in R. K Gupta, Klaus Bosselmann and Prasenjit Maiti (eds.), *Global Environment: Problems and Policies* (New Delhi: Atlantic Publishers & Distributors (P) Ltd., 2008), v. 3, pp. 59-60.
12. Alex de Sherbinin, Koko Warner and Charles Ehrhart, "Casualties of Climate Change: Sea-level Rises Could Displace Tens of Millions," *Scientific American*, January 10, 2011, see http://www.scientificamerican.com/article.cfm?id=casualties-of-climate-change, accessed on June 18, 2011.
13. n. 11, p. 63.

14. "Siachen has Receded 800 m in 20 Years: UN," *The Indian Express* (August 19, 2009), see http://www.indianexpress.com/news/siachen-has-receded-800-m-in-20-years-un/503886/0, accessed on May 10, 2011.
15. Nitin Pai, "Climate Change and National Security: Preparing India for New Conflict Scenarios," *Pragati - The Indian National Interest Review* (New Delhi: April, 2008), no. 1, p. 2, see http://nationalinterest.in/wp-content/uploads/2008/04/inipolicybrief-no1-climatechangeandnationalsecurity-nitinpai-april2008.pdf, accessed on February 10, 2011.
16. See, Nadeem Iqbal, "Economic Threat may Push Pakistan to go Nuclear," *The Asia Times,* February 6, 2002, http://www.atimes.com/ind-pak/DB06Df02.html, accessed on June 13, 2011.
17. See, Julian Borger, "Darfur Conflict Heralds Era of Wars Triggered by Climate Change, UN Report Warns," *The Guardian,* June 23, 2007, http://www.guardian.co.uk/environment/2007/jun/23/sudan.climatechange, accessed on June 17, 2011.
18. See, for more information, "Security Council, in Statement, Says 'Contextual Information' on Possible Security Implications of Climate Change Important when Climate Impacts Drive Conflict," SC/10332, July 20, 2011, http://www.un.org/news/press/docs/2011/sc10332.doc.htm, accessed on July 20, 2011.
19. See, H. Josef Hebert, "Turmoil from Climate Change Poses Security Risks," http://www.usnews.com/science/articles/2009/10/28/turmoil-from-climate-change-poses-security-risks?s_cid=rss:turmoil-from-climate-change-poses-security-risks, accessed on June 25, 2011.

2

Drivers of India's Climate Change Policy

> "There is now general agreement that the environment cannot be protected by perpetuating the poverty of developing countries. Their basic concern is with development and this is as it should be. But it is also no longer acceptable to take as given that a certain degree of environmental degradation and over-exploitation of natural resources in the cause of promoting growth is inevitable."
>
> — Dr. Manmohan Singh, Prime Minister of India

The issues of climate change and energy management have gained salience in India in the current scenario as the energy requirements of the country have been rising rapidly, and at the same time, the impact of climate change have been worsening across the world. India's fossil fuel consumption has risen dramatically since its Independence, while its alternative energy use has seen only a marginal increase. India's energy consumption is expected to grow further in the years to come, taking into consideration the growing population and the expected growth in the Gross Domestic Product (GDP). However, change in India's energy strategy based on the 'green economy' concept that started in the last decade of the 20th century is yet to gain sufficient momentum. On the one hand, the Indian establishment believes that fulfilment of energy requirements should be ahead of the responsibility towards environment protection. On the other hand, India strongly contends that sustainable development is entrenched in the Indian ethos. Therefore, it is imperative for India to mainstream climate change to sustainable development. It would help India in designing a strategy to fulfil its energy requirements using alternative sources of energy such as wind, solar, etc.

In the current context, especially with the signing of the Indo-US Nuclear Agreement, nuclear renaissance is being looked upon as the ground-breaking solution for both tackling climate change and ensuring sufficient production of electricity as it is considered one of the cleanest sources of energy. At the same time, the potential of nuclear energy has been questioned by many critics, particularly in the wake of the nuclear crisis in Japan. India has to grapple with the consequences of climate change on the one hand, and the costs of mitigation and adaptation on the other. The question is whether the country should focus on alleviating poverty or on mitigating climate change. The need of the hour is to understand that it is not a question of either-or; it is to synchronise alleviation of poverty and prevention of environmental degradation, and regard development as the most important adaptation strategy; this could happen only if development is sustainable and not consumerist.

Another important point of discussion is that whether India should divert its resources equally to both adaptation and mitigation, considering adaptation is a bigger challenge for any developing country, and that India does not have any international obligation to reduce its emissions as yet. If India adopts and implements a strong policy of mitigation, then the question of lack of financial and technical resources comes up. Therefore, the chain is long, and the stakes are too high for India and the world. India's role has been increasing at the international level, and an issue such as climate change adaptation and mitigation requires India's active participation.

This chapter attempts to analyse the drivers to the steps taken by India to address concerns with regard to the growing challenges faced by it from the escalating CO_2 emissions at the domestic level. India's perceptions of climate change, adaptation and mitigation are centre not only to its domestic actions but also to its international policy. Not only has the chapter made an account of the National Action Plan strategies and their effectiveness, it has also portrayed the specifics of different State Action Plans and their goals briefly. The objective of the chapter is also to draw a comparison between those State Action Plans and highlight certain gaps in the national-level policies. It explores the policies enacted by the Government in two specific sectors—power and transport (the two biggest contributors to GHG emissions)—and their feasibility. The economics of climate change in terms of the environment versus development debate as well as the costs of adaptation and mitigation have also been discussed. Besides the nuclear energy option, the other issues that entail attention of the policy-makers and that are reflected in this chapter are energy efficiency and the dearth of low-cost green technology in the market. In this context, it scrutinises the contribution of the Clean Development Mechanism (CDM) to India's strategies. It has also tried to draw a parallel between India's dependence on import of fossil fuels currently, and its possible dependence on the import of renewable technology in future.

Climate Change and Associated Economics: Centre to the Issue of Energy Mix

Climate Change and Energy Management: 'Two Sides of the Same Coin'

India's climate change policy is heavily contingent on its energy mix. India's energy policy, in turn, is driven by a number of factors including environmental concerns. First, any growing economy can sustain its growth only if it has a reliable supply of energy, usually in the form of electricity, gas and petroleum products. Second, energy supplies are essentially strained by increasing household incomes. Third, over-reliance on imported resources including gas, crude oil, petroleum products and coal is a direct offshoot of negligible domestic reserves of fossil fuels. Fourth, the urge to adopt cleaner and efficient fuels and technologies emanates from the awareness of the environmental impacts of meeting energy requirements. Therefore, any policy that is implemented to mitigate climate change has a direct bearing on the energy policy of the country and vice-versa.

Fossil Fuels: the Economics of Burden

According to an International Energy Agency (IEA) report (2007), coal/peat accounts for nearly 40 per cent of India's total energy consumption, 27 per cent combustible renewable and waste, nearly 24 per cent oil, 6 per cent natural gas, almost 2 per cent hydroelectric power, nearly 1 per cent nuclear, and less than 0.5 per cent other renewable.[1] When talking about India's climate change policy, besides the energy mix, it is equally important to emphasise on its heavy dependence on imports to meet its energy demands for two reasons—first, India's financial resources are being exhausted for buying fossil fuels, which not only augments emissions but also reduces investment in renewable energy Research and Development (R&D), and second, owing to the lack of self-sufficiency, India might end up depending on industrialised countries and other emerging economies for renewable technology in future. India's growth is theoretically and practically, therefore, determined by factors such as international markets, availability of global energy reserves, production capacities and geopolitics. For example, the oil-rich region of West Asia is currently highly politically unstable; this could potentially choke the fulfilment of India's energy requirements particularly due to rising prices. In fact, there are signs of that already. India imports 70 per cent of its oil requirements that consume nearly 80 per cent of its foreign exchange reserves. The chief suppliers of oil for India are Saudi Arabia (23 per cent), Iran (17 per cent), Nigeria (11 per cent), Iraq (10 per cent), Kuwait (9 per cent), the UAE (9 per cent), Malaysia (4 per cent) and Yemen (3 per cent). India also imports coal from Australia, Indonesia and the US for a little over 10 per cent of its requirements. In the natural gas sector, India imports about 15 per cent of its total requirements.[2]

The Renewable Energy Tale

India's renewable energy policy dates back to 1947 when India came out of the clutches of the British Raj especially with its robust hydroelectric power projects. Thereafter, the first Department of Non-Conventional Energy Sources was established in 1982. However, it was only in the 1990s that the policy got reinvigorated largely due to international circumstances. In order to conform to Agenda 2 of the UN Conference on Environment and Development (The Earth Summit), the Government of India began to lay heavy emphasis on identification, exploration and utilisation of renewable resources. It also created a separate ministry for non-conventional energy sources. Among the renewable energy sources, hydroelectric power has an installed capacity of nearly 38,000 MW, wind power of over 12,000 MW and solar power of nearly 18 MW. The share of renewable energy in installed capacity has grown from 9.7 per cent in 2009 to close to 11 per cent in 2010. In addition, the country has in excess of 4 million family-size biogas plants installed. The total projected capacity by the end of 2022 in wind power is 38,500 MW and solar power is 20,000 MW.[3] Most of the policies formulated by the Government of India in the field of renewable sources of energy are contained in the Five Year Plans brought out by the Planning Commission, especially the 11th Five Year Plan (2007-12), and the other plans are enshrined in the Integrated Energy Policy approved by the Planning Commission in 2006 with the broad objective of meeting energy demand 'at the least cost in a technically efficient, economically viable and environmentally sustainable manner'.[4] Though India has set foot on the path to increasing the share of renewable energy in its economy, it has a long way to go primarily due to the lack of infrastructure and public-private partnerships. During interactions with members of the Planning Commission, India's dismal condition in terms of investment in renewable technology came to light. However, the authorities have started to show the right political signals favouring renewable energy which shall be discussed later in the chapter.

Exploring the Nuclear Energy Option

India's advancement in nuclear energy has been remarkable. Despite the imposition of a nuclear apartheid by the international community, India managed to develop its own indigenous nuclear technology and energy network. Whether it could be a solution to the climate change issue or not is debatable, still it has gained salience in a dichotomous setting – environmental change and post-Fukushima. Nuclear power is the fourth largest source of electricity in India. India has become a new entrant and a prospective active participant in the International Thermonuclear Experimental Reactor project along with the Generation IV International Forum. India is the global leader in the development of thorium-based fast breeder reactors. The Indo-US Nuclear Deal, signed in 2008, has paved the way for India to conduct nuclear trade and commerce with 45 countries of the Nuclear Suppliers' group (NSG). India presently has an installed operating capacity of 4.8 GW and an additional 2.7 GW capacity is close to completion. With the setting up of new units, India's capacity is being touted to reach 40 GW by 2022. However, this highly ambitious goal might not be achieved by the slated timeline. According to a research conducted in 2010, seven operating units of Pressurised Heavy Water Reactor (PHWR) would provide an additional capacity of 5 GW, the fast breeder reactors of 6.5 GW, Light Water Reactor (VVER-1000 and VVER-1200) of 6.6 GW, LWR/EPR (Evolutionary Power Reactor/European Pressurised Reactor) of 9 GW, additional LWR of 20-25 GW, Advanced Heavy Water Reactor (AHWR) of 1.2 GW, adding to a total of around 40 GW of nuclear power capacity, which is far below the Government's expectations of 48-63 GW by 2032.[5] Nuclear energy has become one of the major poles of the climate change debate, especially in a country like India, as it is not only being earmarked as a source of energy that could fulfil the electricity demands of a majority of the population, but also as a primary source of mitigation of climate change. It is very clear from the country's establishment's gesture of giving the green light to the controversial Jaitapur plant on the 25th anniversary of the Chernobyl disaster that nuclear energy will certainly be on top of the agenda at least for the next 20 years. Whether or not the country aims to achieve parity with countries such as France in this respect is debatable, it is contingent on a number of factors—not only environmental but also socio-economic. The future of the nuclear renaissance will be discussed further in the fifth chapter.

The Environment versus Development Debate

One aspect of India that needs to be understood lucidly is the basis for the existence of a whole range of poles in the country. This has been touched upon briefly in the first chapter. In this chapter, the attempt would be to analyse it from the perspective of energy requirements as they are the biggest drivers of India's climate change policy. It has been noted that one of the major factors that led to a decline in the Foreign Direct Investment (FDI) is the negative role played by the Ministry of Environment and Forests under the erstwhile minister, Mr. Jairam Ramesh. These sentiments were expressed at Davos 2011, in which a few foreign companies had expressed concerns about the stalling of various projects such as Posco, Vedanta, etc. Some of the Indian politicians also alleged that his actions were partly responsible for a 36 per cent dip in the FDI in April-September 2010.[6] The Prime Minister himself has re-articulated India's age-old stand on this debate by stating that 'poverty is the biggest polluter',[7] and that there needs to be a balance. This debate dates back to the days of Indira Gandhi's reign; when India was asked by Ronald Reagan in the early 1980s to 'pull up its boots', she retorted that India was not in a position to pull up its boots as Indians did not have boots.[8] As a case in point, nearly 400 million Indians are reported to be without electricity.[9] Not only is part of the population cut off from the oil network, the country has one of the world's highest taxes on fossil fuels. The social repercussions of development and industrialisation such as displacement, shoddy compensation, etc. therefore have taken precedence over environmental ones. It is this amalgamation of social and environmental issues that has ensured indirect environmental justice to some extent. At the same time, Environmental Impact Assessment (EIA) has also been held responsible for the stalling of most of the developmental projects in the north-east, a part of India that is said to be alienated from the rest of the country primarily due to the lack of development. Dr. Manmohan Singh has asserted time and again that no regulatory framework should take the form of a license permit raj that could take India back to the pre-liberalisation era.[10] Environmental concerns, thus, from being a non-issue, has turned into a 'thorn in the flesh' of the successive Indian Governments.

Therefore, the administrators have to grapple with three issues on each side of the divide while formulating climate change policy. On the one hand, the growing challenges from climate change are immense and the scientific predictions do not seem quite bright either; internationally, the pressure on India is rising to take multiple steps forward to address climate change, and to reach a viable international treaty that would facilitate free flow of adaptation and mitigation funds into India; the dreams of achieving 'sustainable development', which the establishment claims is entrenched in the Indian ethos, are contingent on the country's environment-friendly policies in the current scenario. On the other hand, a significant population of the country still subsists below the poverty line and thus, the rich-poor divide needs to be bridged; the middle class is growing at a rapid pace and their energy demands are escalating at an even faster pace; the so-called 'bane of democracy' as some would call it or 'realpolitik' as others would, requires the political class to be more concerned about the fulfilment of immediate needs than the consequences of the current policies that could unravel themselves much later.

Adaptation and Mitigation of Climate Change: India's Strategy

Stepping Stones to Climate Change Policy-Making

As discussed above, the policies that India has adopted in order to address the challenges of climate change synergise the reduction of GHG emissions with the process of pursuing sustainable development. Besides energy efficiency, energy conservation and renewable energy, the Government is also focussing on other policies such as afforestation, wasteland development and fuel substitution. The integration of climate change and sustainable development strategies is institutionalised through the Ministry of New & Renewable Energy, the Bureau of Energy Efficiency (BEE), and the Technology Information, Forecasting & Assessment Council (TIFAC). The other institutions created by the Government of India to not only measure climate change but also to brainstorm solutions include the National Committee to Assess the Impacts of Climate Change and the Prime Minister's Council on Climate Change. The former includes meteorologists, climate modellers, energy economists, representatives of key Ministries, etc. and it is chaired by the Principal Scientific Advisor to the Prime Minister.

Institutionalisation is the first step towards mainstreaming; the next step would undoubtedly be designing the policy framework. This stage has also been marred by several differences of opinion, understanding and interpretation. It is the 'perspectival' number-game that has yet again played a crucial role in the stage of policy-formulation. This chronicle is essentially centred on the dispute between quantum emissions and per capita emissions. So much so that every report released by the Government of India or any government agency, besides making public the policies, also attempts to create a platform for an unending debate on the question of emissions based on figures. The GHG emissions are recorded to have increased at a compounded annual growth rate of 4.6 per cent excluding the agriculture sector.[11] The 2007 GHG estimates also make India the fourth largest emitter[12] and it is expected to become the third largest emitter of energy-related CO_2 by 2015 as per the predictions of the International Energy Agency.[13] A new study shows that India has already become the third highest emitter pushing Russia down to the fourth place. Yet India's per capita CO_2 emissions are approximately 1.38 tonnes per person, compared to a world average of 4.49 tonnes per person.[14] This is the reason why Prime Minister Manmohan Singh has upheld that India's per-capita GHG emissions would not exceed those of developed countries even while pursuing pro-development or pro-growth policies.[15] In fact, according to Dr. Indrani Chandrasekharan, Adviser, Planning Commission, "India's emissions, whether quantum or per capita, can never match the levels of emissions of the developed countries even if the country's growth percentage touches 10 per cent." However, she admits that though India has been successful in bringing down emissions by tinkering with a few sectors through various methods, it needs concrete future plans.[16]

Labyrinth of Perceptions and Incongruities

One of the important distinctions that the Indian establishment has stressed on is the one between climate change and climate resilience. While it clearly lays out the policy for climate resilience (that is governed by natural phenomena mostly) based on meteorological measurements, it lacks expertise on climate change as of now. India stands by the belief that climate change is not just an anthropogenic phenomenon[17] and at the same time, it blames industrialised countries squarely for climate change. In short, India is still in the process of conducting climate change-related research and arriving at convincing conclusions. As far as the understanding of the public about climate change is concerned, a recent global public opinion poll revealed that only 19 per cent of Indians 'believe that global warming is a sufficiently serious and pressing problem to merit taking immediate action as it involves significant cost' in comparison to 42 per cent of Chinese and 43 per cent of Americans.[18] Therefore, the underlying fact is that India is yet to mainstream climate change in its policy in true spirit and letter.

Before going into the specifics of these policies, it is imperative to understand India's position on mitigation and adaptation. The concerned authorities do not see much difference between the two as according to them, both are complementary to each other; any mitigation strategy could also contribute to adaptation efforts and vice-versa. Though the country's focus is more on adaptation as of now, mitigation is not being overlooked. Most of the policies and funds are oriented more towards climate resilience than climate change, that is, strategic knowledge, disaster preparedness and rehabilitation. "Adaptation has to be part of the way we live. We have to make up for the loss and make sure that loss does not happen in future," says Dr. Indrani Chandrasekharan.[19] Yet again, the two biggest factors that determine India's adaptation and mitigation strategies are a booming economy and the increasing number of extreme weather events. But the unfortunate part is that as far as climate change is concerned, there is no sector-wise estimation of the capital that would be required for mitigation purposes. This is also evident in the fact that the State Governments, though allowed to have their own adaptation strategies, have to frame their mitigation strategies in consonance with the Central Government's policies.

Since India does not have an obligation under any international agreement to take actions for mitigation, it needs to focus more on the adaptation front according to a majority of policy-makers. In fact, the defence sector is not even earmarked as one of the sectors in the list of emissions inventory; defence has been identified as one of the biggest energy consumers in the world and every single developed country has been focussing equally on reducing emissions from this. In India, the Ministry of Defence (MoD) does not even have an environmental policy for the country's armed forces in place. Even in the area of adaptation, the country is purportedly in the process of assessing the risks of climate change and the plausible adaptation strategies.

Policies and Prospects Galore

India has undertaken a series of actions and enacted a series of policies, which not only call for poverty alleviation but also ecological sustainability, regulatory and voluntary mechanisms, and efficient strategies among others. The National Action Plan on Climate Change (NAPCC), approved by the Prime Minister on June 30, 2008, is the core of India's climate change policy at the domestic level. Under its umbrella, the Government of India has adopted eight National Missions—National Solar Mission, National Mission for Enhanced Energy Efficiency (NMEEE), National Mission on Sustainable Habitat, National Water Mission, National Mission for Sustaining the Himalayan Ecosystem, National Mission for a Green India,

National Mission for Sustainable Agriculture and National Mission on Strategic Knowledge for Climate Change.

While launching India's NAPCC, Dr. Manmohan Singh stated: "Our vision is to make India's economic development energy-efficient. Over a period of time, we must pioneer a graduated shift from economic activity based on fossil fuels to one based on non-fossil fuels and from reliance on non-renewable and depleting sources of energy to renewable sources of energy. In this strategy, the sun occupies centre-stage, as it should, being literally the original source of all energy. We will pool our scientific, technical and managerial talents, with sufficient financial resources, to develop solar energy as a source of abundant energy to power our economy and to transform the lives of our people. Our success in this endeavour will change the face of India. It would also enable India to help change the destinies of people around the world."[20] The most ambitious Mission that the Government of India has launched in the past few years is the Jawaharlal Nehru National Solar Mission (2010) under the brand name, 'Solar India', which intends to deploy 20,000 MW of solar power by 2020 through a 3-phase approach.

Some of the states have also introduced State-specific Action Plans on Climate Change (SAPCC). The Delhi Plan (2009-12) gives an idea of what the Delhi Government has done and what it will do in future to complement the steps being taken by the Central Government under the aegis of the NAPCC. It takes account of the green agenda of the State, especially in the areas of air pollution, water, noise, municipal waste and greening. For instance, it highlights the setting up of an Air Ambience Fund that promotes battery-operated vehicles. It lays heavy emphasis on the closure of all major coal-based thermal power plants in and around the capital. One of the concerns of the Delhi Government has been the paucity of the supply of natural gas that could come in the way of the functioning of natural gas-based power plants. The Government could struggle to meet the power demand which is increasing at the rate of 10 per cent annually[21] in future due to not only the inability of the newly commissioned plants to produce enough power, but also the unwarranted delays in the commissioning of plants at Chandrapura, Mejia and Maithon. The Commonwealth Games in 2010 had compelled the Delhi Government to shut down some of the plants in order to reduce the pollution levels of the city; for example, the Rajghat

Power House was shut down temporarily, but the authorities resumed its operation as the total output from the new plants could not meet the demand. The promotion of solar heating devices, Compact Fluorescent Lamps (CFLs), Compressed Natural Gas (CNG) and bio-fuel, greening, waste-water treatment and reuse, generation of electricity from municipal waste, and environmental education and training have been initiated by various governmental agencies.[22]

Gujarat is one of those states which has even allocated a budget to invest substantially in the Clinton Climate Initiative for setting up a 3,000 MW Solar Park, other green solar projects, Jyotigram Yojana to reduce electricity transmission losses, Bus Rapid Transit System which has already received accolades worldwide for its services in Ahmedabad, and is now being planned in Surat and Rajkot, a state-wide drinking water grid, planting of mangroves etc. It is also the first State to form a separate climate change department.[23] Therefore, like the NAPCC, the SAPCCs have also largely been a mixed bag. At the outset, these plans seem to be restricted to low-hanging fruits which cannot be effective in the long-run and could be too little even in the present. One needs to go deeper into the impact of these actions, such as the impact of the metro transport system on the number of private vehicles on the roads of Delhi or the impact of the closure of coal-based power plants on the Government's policy of regulation of electricity consumption patterns of the population, and accordingly draw estimates of what could be in store for India in future, with and without large-scale use of green technology across the spectrum.

The Government has taken several steps to remove barriers that earlier restricted private players from exploring, extracting, converting, transmitting and distributing primary and secondary energy. Some of them include Electricity Act 2005, Tariff Policy 2003 and Petroleum & Natural Gas Regulatory Board Act 2006. Also, the promotion of use of renewable energy sources was reinforced by the introduction of feed-in-tariffs in 2008, Rural Electrification Policy 2006 and the New and Renewable Energy Policy 2005. Due to the easing of regulations, India has emerged as 'one of the top performing clean energy economies in the twenty-first century' with an investment of US$ 10.2 billion in clean energy, placing it at the sixth spot among the G-20 countries in 2012. Blending of ethanol with gasoline is not only being encouraged, it is also authorised across the country as evidenced by the Government's decision in 2003 to authorise a 5 per cent blending of ethanol with gasoline in 9 states and 4 Union Territories in 2003, which was extended to 20 States and 8 Union Territories in 2006.[24]

Use of biofuels not only saves petrol but also results in lower emission of pollutants. On the flip side, there are reports which indicate a certain rollback on the part of the Indian Government on the front of bio-fuels, especially jatropha due to the lower production of jatropha seeds, rising food prices, fear of diversion of land for bio-fuels and its impact on food production, and the purported negative consequences of first generation bio-fuels on the environment. Based on these reasons, the erstwhile Minister of Environment and Forests, Mr. Jairam Ramesh rejected Executive Secretary, Christiana Figueres' suggestion that India should 'explore business possibilities of jatropha plantations for bio-fuels' with private bio-fuel companies backed by the International Air Transport Association (IATA) in 2011.[25]

In addition, the Government has also taken steps to bolster disaster management efforts especially in terms of response to extreme weather events, such as infrastructure design, communication networks, and protection of coastal areas, health care services and capacity building both at the regional as well as state levels. The Government has been pumping funds into the development of arid-land crops and pest management; and drought proofing programmes by which the adverse effects of drought on production of crops and livestock could be curtailed; these are the Crop Insurance Scheme and the Credit Support Mechanism that insures farmers against climate risks and facilitates extension of credit to farmers respectively.

A Brief Sectoral Policy Study: Power and Transport

In India, the coal-driven power sector is the highest contributor to GHG emissions. Therefore, reducing the carbon footprint of the power sector has been one of the chief areas of focus for the Government of India. One of the most recent steps taken by the Government in this regard was the launch of the Perform-Achieve-Trade (PAT) mechanism under the approval of the BEE in January 2011. Through this mechanism, energy-intensive large industries would be assisted in the process of reducing energy consumption under the Energy Conservation Act 2001. This would be accomplished by allowing industries that consume less energy to sell efficiency certificates to the non-efficient ones under the mechanism. The Ministry of Power (MoP) has 'notified industrial units and other establishments consuming energy more than the threshold in 9 industrial sectors—Thermal Power Plants, Fertiliser, Cement, Pulp and Paper, Textiles, Chlor-Alkali, Iron & Steel, Aluminium and Railways in March 2007 as Designated Consumers under the PAT scheme'.[26] According to the reports, "By adopting the mechanism they would be saving to the tune of about 10 million tonnes oil equivalent in the next three years."[27] One of the other noteworthy measures taken by the Government to promote energy efficiency is its decision to invest in coal-fired Ultra-Mega Power Plants with a power generating capacity of 4 GW per unit. The Government has proposed to convert some of the most polluting thermal power plants into gas plants to reduce the carbon footprint such as the one in Badarpur, Delhi. India's first super critical power plant (660 MW) was set up at Mundra, Gujarat. This technology helps in reducing emissions and increasing efficiency.

Under the auspices of the NMEEE (one of the missions of the NAPCC), the Government expects to save 10,000 MW by the end of the 11th Five Year Plan in 2012 through a series of activities and institutions such as the PAT Mechanism, Market Transformation for Energy Efficiency (shift to energy efficient appliances), Financing of Energy Efficiency (to promote carbon finance and other mechanisms that would facilitate capturing of future energy savings), Power Sector Technology Strategy, Energy Efficiency Services Ltd, BEE as well as creating awareness programmes in schools, industry, commercial, agriculture and domestic sectors.[28]

In the transportation sector, one of the other biggest emitters, the Government has begun a transition process to cleaner fuels like CNG in the capital (Delhi) and bio-fuel that is being currently used by Karnataka State Road Transport Corporation (KSRTC) buses. As part of its Biofuels Policy, the Indian Government has set the target of meeting up to 59 per cent of the country's transport demand through greener fuels by 2020.[29] It has also been concentrating on effective public transport systems such as metro trains for large metropolitan centres under the auspices of the National Urban Transport Policy (2006) and National Urban Renewal Mission. The National Auto Fuel Policy (2003) necessitated all new four-wheeler vehicles in eleven cities to meet Bharat Stage III emission norms for conventional air pollutants and comply with Euro IV standards (Bharat Stage IV) by 2010.[30] These measures are yet to encompass the whole country; the progress has been slow. For instance, introduction of metro transport has been successful in Delhi, but in other cities such as Mumbai and Bengaluru, the pace at which the project has evolved has been glacial. When it comes to vehicular emissions, one also has to keep in mind the rapid increase in the number of private vehicles in the country which remains uncontrollable. Therefore, even if the public transport network is enhanced, the number of vehicles will continue to grow and the end result only would be a shift of population from one branch of public transport to another depending on costs and comforts.

Future as Viewed through the Present

One of the biggest steps that India has taken in the post-Copenhagen scenario is to form an Expert Group on a Low Carbon Strategy for Inclusive Growth. The interim report of the Group under the Chairmanship of Dr. Kirit Parikh has come up with a series of options that could reduce India's emission intensity over a particular time frame. According to the report, the Government could bring down the emission intensity of India's GDP by 23 to 25 per cent over the 2005 levels if it adopts 'Determined Efforts' and by 33 to 35 per cent over the 2005 levels if it adopts 'Aggressive Efforts' by focussing on the sectors of power, transport, industry, buildings and forestry.[31] While the former stresses on the strengthening and upgradation of existing policies and the ones are in the pipeline, the latter emphasises on the introduction and implementation of new policies. Apart from the fact that the report fails to deviate from the 'business-as-usual', it also does not look forward to decrease the electricity demand of the country. Dr. Kirit Parikh, on the other hand, defends the report by saying that it emphasises on moving away from the 'business-as-usual' scenario; however, the fulfilment of ambitious targets requires a huge amount of money.[32] It is extremely important for the Planning Commission to retrace steps to its Integrated

Energy Policy report that gave equal importance to both increasing energy efficiency as well as decreasing energy intensity. India's policy-makers promote energy efficiency as a short-term measure in order to bring down energy intensity. They deem energy efficiency as the way forward and therefore, the key is to be as efficient as possible through technical progress and innovation. As Dr. Parikh observes, "Energy efficiency makes activities more profitable; the economy grows faster and the energy consumption increases. However, the energy intensity decreases."[33]

There is a section of economists according to whom energy efficiency reduces demand in the course of time. It is believed that energy efficiency essentially leads to reduction in wasted energy besides the fact that consumers would be expected to use lesser amounts of energy to achieve the same level of utility. There are two sides to this debate—first, according to the 'Rebound Effect',[34] as the demand and the cost of using the particular resource decreases, there would be a rebound in demand, and second, according to the 'Piggy Principle',[35] consumers would rather try to achieve a higher utility at the same expense. Moreover, there is a limit to which energy efficiency can be increased and the energy demand would only increase with the increasing population.

Unfortunately, even the data presented in the report has come under the scanner for various reasons. First, the data generators are mostly government statistical agencies; however, the estimates are generated by non-governmental agencies that may or may not be cross-checked before being used for framing policies. These non-governmental agencies could even be influenced by external entities in terms of financial and political agencies that could be detrimental to the country's interests. Second, as a report of the Centre for Science and Environment reveals, the data is 'weak, outdated and not referenced'. The review also concludes that the projected 2020 performance 'has already been achieved or will be achieved soon' in some sectors while in others, the projected 2020 performance is 'overly ambitious and will be difficult to achieve'.[36] Yet, it is also necessary to acknowledge that India is the first non-Annex I country to update its GHG Emissions Inventory. The Planning Commission has also defended itself by stating that the figures cannot be inaccurate since most of them are an outcome of satellite mapping. The future of climate change policy depends to a great extent on the subsequent reports of the group. The efforts recommended in the report and alternatives would be analysed in a later chapter.

In the past two years, India has also shown commitment to levy a carbon tax that would eventually contribute to a National Clean Energy Fund. Besides, it has also implemented various other measures such as the Bachat Lamp Yojana (CDM Programme of Activity for mass distribution of CFLs), the setting up of a REDD+ (Reducing Emissions from Deforestation and Forest Degradation) Technical Group, and the Solar Energy conclave that discussed the financial and technical aspects of the National Solar Mission for an effective implementation, among others.[37] The reality is hard to gauge with very little transparency on the findings of the conclave since the ambitious targets set by the Government can be achieved only if the investment is increased manifold. The Government's initiative to increase forest cover by 5 per cent has reportedly failed; it could meet only one-third of the target. One could also argue that the Government should instead be commended for 'maintaining' the forest cover despite immense pressure from various quarters.

India and the CDM

Coming to India's international cooperation, it has endorsed the CDM and Technology Transfer from the developed to developing countries. Though these are two different concepts, CDM could itself become a launch pad for technology transfer in future. The reason for the endorsement of technology transfer is primarily based on the grounds of lack of resources to produce green technology in the country. The CDM clearly allows India and other developing countries to proceed with their developmental plans and at the same time, mitigate and adapt to the effects of climate change. Moreover, the CDM forms the bedrock of the developing world's perspective that largely puts the obligation of climate change mitigation and adaptation on developed countries. However, India, as emphasised earlier, would like to see that 'a CDM project leads to a real technology transfer giving the country the ability not only to operate technology but also to replicate and innovate'.[38] This brings to focus the issue of Intellectual Property Rights that is still a bone of contention between the industrialised and non-industrialised countries. Similarly, there are other obstacles such as the additionality clause of the CDM and pricing of technology. Internally, the investment environment and awareness about the CDM are generally poor. Besides, allegations of corruption have been levelled against various entities involved in the mechanism, such as the example of how an Indian wind developer kept the CDM authorities in the dark about the 'lucrative tax credits their projects were earning'.[39]

'The CDM National Designated Authority' has already approved over 1500 CDM projects and figures reveal that out of these, over 600 have been registered at the CDM Executive Board and over 900 are at or have even crossed the validation stage as of March 31, 2011. 27 per cent of the registered projects are in the wind power sector, followed by biomass (26 per cent), hydroelectric power (12 per cent), waste gas/heat utilisation (11 per cent), energy efficiency (11 per cent), cement and fuel switch (3 per cent), biogas and methane avoidance (2 per cent), HFC reduction/avoidance, N_2O decomposition and other renewable energies (1 per cent). They have resulted in the issue of over 93 million tonnes of certified CO_2 emissions reductions.[40] These projects have accelerated to some extent investment in the renewable energy and energy efficient technologies that were earlier viewed sceptically by policy-makers and business enterprises on account of factors of unpredictability and unreliability.

Costs of Mitigation of Climate Change for India

The impact of CO_2 emission reduction strategies will come at a cost if India does not change its strategy in a manner that balances the growing population and the escalating needs of its population. Studies show that it could even lead to lower GDP and higher poverty. As per a study, "A 30 per cent CO_2 reduction over a period of 30 years using annual emissions reduction targets leads to a fall in GDP of 4 per cent and raises the number of poor by 17.5 per cent in the 30th year (that is, if 2000 be taken as the baseline, these changes would occur by 2030)." If a cumulative emission reduction target of 30 per cent is taken into consideration, then the fall in GDP would be 1.4 per cent and the increase in the number of poor would be 6 per cent by the 30th year. Moreover, the compensation for the loss in welfare would require immense capital inflows (US$ 278 billion for annual reduction and US$ 87 billion for cumulative reduction). However, it has been found that tradable emission quotas could make India carbon-efficient as India could transform itself from a net seller in the first 25 years to a net buyer in the later years.[41] There have been attempts on the part of various scholars to study the long-term GHG emissions trends and mitigation costs. According to the ERB (Edmonds, Reilly and Barns) model, India would continue to be a net energy importer during the next century due to limited domestic supply and increasing energy demand. The share of India's contribution to global emissions could rise up to 10 per cent if no mitigation strategies or technical/technological interventions are adopted. The same model uses carbon tax as a mitigation strategy and concludes that 'a severe tax regime' could result in a Gross National Product (GNP) loss of almost 3.5 per cent, which is expected to stabilise later on; plus it could cost the exchequer annually US$ 60 billion in 2050.[42]

The best way to deal with the costs is to enhance the role of the CDM and Technology Transfer. India could also contemplate emissions trading as one of the options towards achieving obligations towards reducing GHG emissions. It is considered one of the most cost-effective incentive-based approaches; however, a national-level emissions trading mechanism can only sustain itself if it is linked to the global/regional emissions trading mechanism. Also, there is a fear that it could get converted into a license raj system wherein the Government intrudes in the system not only in terms of setting the standards and monitoring, but also in giving permit to the industries. At the same time, there is apprehension among the policy-makers about the benefits of the carbon trading system for developing countries as it is contingent to a large extent on the competitiveness of the market, existence of future markets and the mechanism by which carbon is bilaterally traded in a project-by-project basis as envisioned under the CDM. These mechanisms and approaches would be touched upon in the chapter dealing with the future course that India could take to address climate change. One could say confidently that the path towards a low-carbon economy is daunting for a country such as India unless appropriate burden-sharing rules between the developed and developing countries are framed. The whole endeavour could become a win-win situation if India participates productively in the global GHG mitigation regime. On one hand, the country has projected itself internationally as a country with soaring energy demands, and sometimes even as an energy-starved one that simply cannot afford to stop its growth process. On the other hand, it has been admitted by some of the eminent members of the Planning Commission that there are states with surplus energy and it is only a question of temporary storage that could be a solution to the energy requirements. One has to remember that the benefits of mitigation outweigh the costs of mitigation in the long-run in every way as proved by a study conducted by the Potsdam Institute for Climate Impact research. It says that if the European Union reduces its CO_2 emissions by 30 per cent instead of 20 per cent by 2020, 'up to 6 million jobs would be created; average economic growth would increase by 0.6 per cent annually; in Germany alone, unemployment would fall by almost 3 per cent'.[43] It should also be acknowledged that using this in the context of a single country would not be appropriate since global initiatives would be necessary to reach that stage.

The Saga Continues...

As far as India's energy management is concerned, India is undoubtedly dependent on fossil fuels, especially coal, to meet its energy requirements. Since India's domestic production is limited, it has to depend on imports. This could only affect India's foreign exchange reserves and its capacity to develop green technologies. India's fossil fuel consumption has increased manifold over the years, and it is high time that the country shifts focus to renewable energy sources so that the next generation does not become a victim of the current generation's 'unsustainable' policies. Moreover, India's geography allows it to utilise various renewable sources of energy such as solar, wind and hydro. On paper, it is the lack of resources and the increasing energy demand of an increasing population that have been the biggest stumbling blocks in renewable energy and technology investment. More than these two factors, it is the lack of political will that has retarded the process even though the authorities have shown some propitious signals.

Besides, the paranoia surrounding privatisation and autonomy has also led to the creation of a vacuum in the arena which the Government is struggling to fill.

The 'nuclear lobby' in India has managed to convince the Government that nuclear energy is the only way ahead for the country. This claim could not be completely discarded taking into consideration the steady pace at which fossil fuel deposits are getting depleted, the population is growing and energy demands are escalating. The costs of adaptation and mitigation are expected to be reasonably high, especially for developing countries like India. When it comes to the achievement of long-term sustainable development, one has to remember that the benefits of adaptation and mitigation, not only in terms of climate change but environmental degradation in general, far outweigh the costs. Relocation of a nation state's critical infrastructure itself could ad infinitum exhaust its financial reserves leaving other development programmes unfinished or crippled. India also has to get itself out of the labyrinth of perceptions and definitions in order to boost its ethos of sustainable development rather than replicating the Western model of economic development blindly.

When the going gets tough, the tough get going. This cliched statement has been an inspiration for many a great personality. India is on the right track in what appears to be the Asian century as North America and Europe have already plunged into deep economic crisis. If India wishes to be a global power in the making as labelled by many analysts, it has to show more responsibility and character. Reports disclose that in 2010, developing countries invested more in renewable energy than developed countries, especially Brazil and China. South Africa is also getting ready to launch its South African Renewable Initiative (SARI) despite financing concerns and other obstacles.[44] Therefore, it is imperative for India to pull up its socks and increase its renewable energy investment so that it does not lag behind the rest of the emerging economies in the world. The policy-makers of the country have to realise that not all mitigation policies would be to the detriment of growth and poverty alleviation. Public transport is a classic example. Despite having a huge and flourishing IT sector, very few innovations have been made on the front of renewable energy technology in the country. When there are no dearth of IT brains in the country, nothing should stop the Government from taking advantage of them and find long-term solutions. India and other developing countries have to be sensitive to the environment despite their developmental agenda based on poverty alleviation; at the same time compliance to any unfair international treaty should be dealt with diplomatically to ensure universal access to energy and technology which is key to 'climate justice'—one of the key drivers of India's climate change policy. The policies framed by India tend to focus more on the low-hanging fruits and at some level, the loopholes in them could be exploited by the 'growth-first stonewallers' to thwart the efforts of the 'progressive realists' and the 'progressive internationalists'. These phrases used by an eminent Indian expert, Navroz K. Dubash, will be dissected from all angles in the next chapter as India's global climate change policy is reviewed.

Notes

1. Energy Information Administration, "India Energy Data, Statistics and Analysis – Oil, Gas, Electricity, Coal," *Country Analysis Briefs* (USA: August, 2010), p. 1, see http://ei-01.eia.doe.gov/emeu/cabs/India/Background.html, accessed on February 3, 2011.
2. Narsi, "India's (Fossil) Energy Security Strategy," *Energy Alternatives India* September 4, 2010, see http://eai.in/blog/2010/09/indias-fossil-energy-security-strategy.html, accessed on February 3, 2011.
3. See, "Renewable Energy in India: Progress, Vision and Strategy," Ministry of New and Renewable Energy, Government of India, pp. 8, 14, http://www.mnre.gov.in/pdf/mnre-paper-direc2010-25102010.pdf, accessed on July 15, 2011.
4. See, "Climate Change Mitigation Measures in India," International Brief 2, September, 2008, http://www.pewclimate.org/docUploads/India-FactSheet-09-08.pdf, accessed on February 23, 2011.
5. Rajan Gupta, Harihar Shankar and Sunjoy Joshi, "Development, Energy Security and Climate Security: India's Converging Goals," *Observer Research Foundation* (New Delhi: 2010), pp. 11-13, see http://cnls.lanl.gov/~rajan/Gupta_orf_writeup_v9.pdf, accessed on February 2, 2011.
6. See, "Davos: Development vs the Environment Debate," *NDTV 24x7* February 28, 2011, http://www.ndtv.com/video/player/ndtv-special-ndtv-24x7/davos-development-vs-the-environment-debate/189398, accessed on March 10, 2011.
7. See, "Action Louder than Words," *The Hindustan Times,* June 30, 2011, http://www.hindustantimes.com/StoryPage/Print/715796.aspx, accessed on July 20, 2011.

8. Interaction with Air Commodore Jasjit Singh AVSM VrC VM (Retd), Centre for Air Power Studies on July 6, 2011.
9. See, Jeremy Page, "India Looks to the Sun for Ambitious Surge in Green Power," *The Times,* August 3, 2009, http://www.timesonline.co.uk/tol/news/world/asia/article6736726.ece, accessed on June 15, 2011.
10. See, Manmohan Singh, "We Need a Regulatory Framework that does not Become License Permit Raj," *The Indian Express* (Delhi: July 26, 2011), p. 11.
11. "Interim Report of the Expert Group on Low Carbon Strategies for Inclusive Growth," *Planning Commission, Government of India,* May, 2011, p. 23, see http://planningcommission.nic.in/reports/genrep/Inter_Exp.pdf, accessed on June 20, 2011.
12. See, "India: Emissions up 58%," *India Climate Portal,* May, 2010, http://www.indiaclimateportal.org/component/option,com_policybrief/view,policybriefdetail/id,24, accessed on June 15, 2011.
13. See for more information, International Energy Agency, "World Energy Outlook 2007: China and India Insights," p. 3, http://www.iea.org/textbase/nppdf/free/2007/weo_2007.pdf, accessed on February 07, 2011.
14. "World Carbon Dioxide Emissions Data by Country: China Speeds Ahead of the Rest," *The Guardian,* January 31, 2011, see http://www.guardian.co.uk/news/datablog/2011/jan/31/world-carbon-dioxide-emissions-country-data-co2, accessed on February 19, 2012.
15. Ministry of Environment & Forests, Ministry of Power and Bureau of Energy Efficiency, "India: Addressing Energy Security And Climate Change, October, 2007, pp. 1-2, see for more information, http://www.moef.nic.in/divisions/ccd/Addressing_CC_09-10-07.pdf, accessed on February 03, 2011.
16. Interaction with Dr. Indrani Chandrasekharan, Adviser, Planning Commission, at Yojana Bhawan, New Delhi, on July 13, 2011.
17. Ibid.
18. John Podesta and Peter Ogden, "Security Implications of Climate Change 1: Expected Climate Change over the Next Thirty Years," in Kurt M. Campbell (ed.), *Climatic Cataclysm: The Foreign Policy and National Security Implications of Climate Change* (Massachusetts: The Brookings Institution, 2008), p. 100.
19. Interaction with Dr. Indrani Chandrasekharan, Adviser, Planning Commission, at Yojana Bhawan, New Delhi, on July 13, 2011.
20. Jawaharlal Nehru National Solar Mission: Towards Building SOLAR INDIA, see for more information, http://mnre.gov.in/pdf/mission-document-JNNSM.pdf, accessed on February 03, 2011.
21. See, "Power Shortage: Scorching Summer Ahead for Delhi," *The Indian Express,* March 11, 2011, http://www.indianexpress.com/news/power-shortage-scorching-

summer-ahead-for-d/760922/, accessed on August 08, 2011.

22. See, "Climate Change Agenda for Delhi 2009-2012," http://publicity.delhigovt.nic.in/public/eng/publication/climateChange0003.aspx, accessed on June 09, 2011.
23. See, "Climate Change Action and Adaptation," Climate Change Department, Government of Gujarat, http://moef.nic.in/downloads/others/States-SAPCC-gujarat.pdf, accessed on July 19, 2011.
24. See, "Why India must Push for Ethanol Blended Petrol," *Rediff Business,* July 27, 2010, http://business.rediff.com/slide-show/2010/jul/27/slide-show-1-why-india-must-push-for-ethanol-blended-petrol.htm, accessed on July 27, 2011.
25. Nitin Sethi, "UN Official Lobbied for Private Biofuel Firms during India Visit," *The Times of India,* July 2 , 2011, see http://articles.timesofindia.indiatimes.com/2011-07-02/developmental-issues/29729976_1_bio-fuels-jatropha-iata, accessed on August 20, 2011.
26. Sanjay Dube, Ritesh Awasthi and Vivek Dhariwal, "Can the Learning's from International Examples Make the 'Perform Achieve and Trade (PAT) Scheme' Perform Better for India," *Emergent Versus International* (Gurgaon: March, 2011), see http://www.emergent-ventures.com/UploadedFiles/Videos/A%20Discussion%20Paper%20on%20India's%20Perform%20%20Achieve%20and%20Trade%20(PAT)%20Scheme.pdf, accessed on August 5, 2011.
27. See, Vijay C. Roy, "BEE to Launch PAT Mechanism by April 1," http://www.business-standard.com/india/news/bee-to-launch-pat-mechanism-by-april-1/423342/, accessed on February 03, 2011.
28. Government of India, Prime Minister's Council on Climate Change, National Action Plan on Climate Change, p. 3, see for more information, http://pmindia.nic.in/Pg01-52.pdf, accessed on February 3, 2011.
29. See, "Advanced Biofuel to Meet over 50% of India's Transport Demand," *Biofuels Digest,* May 26, 2011, http://biofuelsdigest.com/asia/2011/05/26/advanced-biofuel-to-meet-over-50-of-indias-transport-demand/, accessed on August 3, 2011.
30. See for more information, "Vehicular Exhaust," http://cpcb.nic.in/Vehicular_Exhaust.php, accessed on August 9, 2011.
31. See for more information, "Low Carbon Strategies for Inclusive Growth: An Interim Report," Planning Commission, Government of India, May 2011, p. ii.
32. Interaction with Dr. Kirit Parikh, Chairman, Expert Group on Low Carbon Strategies for Inclusive Growth, at Integrated Research and Action for Development, New Delhi, on July 19, 2011.
33. Interaction with Dr. Kirit Parikh, Chairman, Expert Group on Low Carbon Strategies for Inclusive Growth, at Integrated Research and Action for Development, New Delhi, on July 19, 2011.
34. Frank Gottron, "Energy Efficiency and the Rebound Effect: Does Increasing

Efficiency Decrease Demand?" *CRS Report for Congress* (Washington, D. C.: July 30, 2001), see http://www.ncseonline.org/nle/crsreports/energy/eng-80.cfm?&CFID=11262148&CFTOKEN=7028302, accessed on August 19, 2011.

35. Marco Bertoli, "Energy Efficiency and its Impact on Energy Demand," *OilPrice.com,* February 28, 2010, see http://oilprice.com/Energy/Energy-General/Energy-Efficiency-and-its-Impact-on-Energy-Demand.html, accessed on August 19, 2011.
36. "Comments of the Centre for Science and Environment on Interim Report of Expert Group on Low Carbon Strategies for Inclusive Growth by the Planning Commission of India," May 06, 2011, p. 1, see http://www.cseindia.org/sites/default/files/Comments%20of%20the%20Centre%20for%20Science %20and%20Environment%20on%20Interim%20Report%20of%20Expert%20Group%20on%20Low%20Carbon%20Strategies%20for%20Inclusive%20Growth.pdf, accessed on July 13, 2011.
37. See, "India: Taking on Climate Change, Post-Copenhagen Domestic Actions," *Ministry of Environment and Forests,* June 30, 2010, http://moef.nic.in/downloads/public-information/India%20Taking%20on%20Climate%20Change.pdf, accessed on July 19, 2011.
38. Jyoti K. Parikh and Kirit Parikh, "Climate Change: India's Perceptions, Positions, Policies and Possibilities," *OECD* (Paris: 2002), p. 22, see http://www.oecd.org/dataoecd/22/16/1934784.pdf, accessed on February 03, 2011.
39. Patrick McCully, "Discredited Strategy," *guardian.co.uk,* May 21, 2008, see http://www.guardian.co.uk/environment/2008/may/21/environment.carbontrading, accessed on July 28, 2011.
40. "Clean Development Mechanism (CDM) in India," see for more information, http://enviroscope.iges.or.jp/modules/envirolib/upload/984/attach/india_final.pdf, accessed on July 28, 2011.
41. n. 37, p. 7.
42. P. R. Shukla and Ashish Rana, "Energy Economy Model Applications: Long-Term Greenhouse Gas Emissions Trends and Mitigation Costs," in Michael A. Toman, Ujjayant Chakravorty and Shreekant Gupta (eds.), *India and Global Climate Change: Perspectives on Economics and Policy from a Developing Country* (USA: Resources for the Future, Washington DC, 2003), p. 229.
43. See, "Environmental Protection = Economic Spurt," *GEO,* (India: July 2011), p. 22.
44. See, United Nations Environment Programme, "Global Investments in Green Energy up Nearly a Third to US$ 211 billion," http://www.unep.org/newscentre/Default.aspx?DocumentID=2647&ArticleID=8805&l=en, accessed on July 28, 2011.

3

The Journey of International Negotiations: Conflict and Cooperation

> "The issue is not only the money. Some people think only funds or money would solve the problems. It did not solve the problems; it will not solve the problems of the present and will not solve the problems for the future."
>
> — Luiz Inacio Lula Da Silva, former Brazilian President

One of the questions that require to be answered is whether climate change should be dealt with as an 'international issue' or a 'transnational issue'. As long as it is treated as an 'international issue', it will continue to be defined and dictated by the nation states. It is also true that every issue that is of global concern is a 'transnational' issue; however, in this domain, the agenda is restricted to debate and discussion. On the one hand, there is a need to strengthen the transnational realm to facilitate the trickle-down effect by which the environmental policies at the national level have an impact as well as are emulated by the grassroots. On the other hand, the need to give more legitimacy to the international regime is also gaining relevance. International climate change negotiations began with the objective of rendering a new dimension to the concept of development, which is sustainability, so that the interests of both the developed and developing countries are complied with. Simultaneously, equal emphasis was given to reduction of emissions, primarily CO_2 emissions. The internationalisation of the issue was a welcome step as the 'coalition of the willing' principle could only jeopardise the situation further. The carbonisation of the debate has led to the focus being shifted to one particular gas (CO_2), and this has indeed turned climate change into a one-dimensional issue. The harshest critics would go one step ahead to allege that this has reduced the issue from a possible catastrophic threat to a profit-making venture.

The journey that began at Stockholm resulted in the creation of a Framework Convention that came into effect after a series of debates across various formal and informal conferences. India was an integral part of them right from the late eighties but its views were never given due recognition. The United Nations Framework Convention on Climate Change (UNFCCC) is the only legally acknowledged international forum for international decision-making on climate change. It symbolises cooperation and consensus. The two-dimensional debate of conflict and cooperation over a majority of international issues has been a hallmark of climate change negotiations as well. On the one hand, nation states with wide-ranging economies at varying levels of development have set aside their political differences in order to address a common issue. The realisation that the principle of 'collective defence' has to be employed in such a manner that every single stakeholder has to wage a war not against an external enemy (that is the environment) but against the enemy within (the nation state itself) has supplanted climate scepticism. The world has come a long way from being in a state of denial about the impact of human activities on the environment to signing treaties of international consequence to prevent colossal destruction due to climate change. Yet, a large section of environmentalists staunchly believes that the UNFCCC has become a territory in itself that has turned into a source of conflict rather than cooperation. The cooperation that the world has witnessed over the years on this issue has been superficial and most often too-little-too-late. The divergences within the developed bloc have given way to the ones between the developed and developing countries. The discourse of political economy has also ensured that until and unless economic gains and losses are shared equally (in the relative sense) by every stakeholder, cooperation would remain ostensible. It would be restricted to consistent and coherent rhetoric with very little action and more emphasis on parlance and political correctness. Many a time, this rhetoric steals the limelight while the real issues get sidelined.

A clear understanding of the evolution of the developed as well as developing countries' positions including that of India at the global climate change negotiations since the inception of the UNFCCC in 1992, especially at Copenhagen (2009) and Cancun (2010) becomes crucial. The major fault-lines guided by multitude of factors including 'common but differentiated responsibility', bottom-up approach versus top-down approach, Environmental Kuznets Curve (EKC) and climate justice among others need to be dissected. These factors have played a decisive role in the formulation of India's perceptions and policies of climate change as well. This chapter attempts to analyse the merits and demerits of the frameworks of the Kyoto Protocol, CDM, and climate financing endorsed by the member nations of the UNFCCC at Copenhagen in terms of implementation.

It also scrutinises the phraseology of both the Copenhagen Accord as well as the Cancun Agreement. This is necessary in the endeavour of articulating the climate change debate in its entirety. It would continue to throw light on India's policy-making expedition by connecting the domestic drivers (discussed elaborately in the second chapter) with its international climate change policy. The chapter seeks to probe the shift in global climate change in politics with the rise of the BASIC group (Brazil, South Africa, India and China) particularly the Asian giants and the emergence of G-2 (China and the US); the subsequent assertion of the Least Developed Countries (LDCs) and the Alliance of Small Island States (AOSIS) and the split in G-77 (a coalition of developing countries) are some of the other developments that require analysis. It would be a worthwhile attempt to examine the steps that have been taken so far by the international political community to address the concerns emanating from global warming at the climate change negotiations.

UNFCCC: A Divided World

Genesis and Nemesis

It was at the UN Conference on Human Environment (Stockholm Summit) in 1972 that the first UN Summit on environment was held. For the first time ever in history, the issue of Chlorofluorocarbons (CFCs) and the creation of an environment ministry were discussed. At this multilateral forum, the link between environment and development was established stressing on the importance of sustainable development; international agreements on ocean dumping, pollution from ships and endangered species trade were also finalised. In the First Climate Conference at Geneva, scientists from varied disciplines finally concurred that a wide range of human activities could indeed contribute to global warming, which eventually led to the establishment of the World Climate Programme and World Climate Research Programme in the form of the Intergovernmental Panel on Climate Change (IPCC). The scientific consensus gained momentum during the 1980s and 1990s at the Villiach Conference (1985), the Toronto Conference (1988) and the Second World Climate Conference (1990) that called for policy response including a 20 per cent cut in carbon emissions by 2005. In addition to these inter-governmental conferences, the European Community and European Free Trade Association members initiated unilateral commitments to stabilise carbon emissions.

All these developments led to the agreement in the form of the UNFCCC at the 1992 Rio Summit (United Nations Conference on Environment and Development). Held from June 03-14, 1992, the Earth Summit discussed the stabilisation of greenhouse gas (GHG) concentrations in the atmosphere, but the treaty was legally non-binding. In addition, the parties to the UNFCCC were classified as Annex I countries (industrialised countries that have an obligation to limit anthropogenic emissions of GHGs), Annex II countries (developed countries that have agreed to pay for the costs of developing countries) and Developing countries (that can volunteer to become Annex I countries under the precondition of achieving significant levels of development). In conjunction with this classification, developed countries were assigned a 'voluntary goal of returning GHG emissions to 1990 levels' while developing countries under the auspices of 'common but differentiated responsibilities' were entitled to receive financial and technological resources from developed countries to adapt to climate change.[1] Another major highlight of the Summit was the creation of a fund under the aegis of the Global Environment Facility (GEF). This Convention was initially signed by 155 countries and the EU and entered into force on March 21, 1994. Currently it has 195 parties—194 countries and 1 regional organisation. The classification of countries sowed the seeds of divisions within the UNFCCC, with some countries obligated to do way more than the rest owing to their greater share of responsibility for the problem. The objective of the conference was to bring the world together to address a global threat; instead, it initiated an eternal clash of interests between the nation states.

Who has the right to develop?

The inter-governmental negotiations with regard to climate change have been plagued by a number of interrelated fault-lines as they involve the political and economic structures of every country that is part of the negotiations. The first and the most important conflict is the one that exists between what the postcolonial experts call, the North and the South, or the industrialised and non-industrialised countries. The two groups differ mainly on the issues of sovereignty and the 'common but differentiated responsibility'. The developing countries' sovereign right to use their own natural resources was diluted to some extent by including this principle in the Preamble and not in the Principles section. As Mathew Paterson points out, "The North felt that putting clauses in the Principles section would create precedents and potentially create new general principles of international law."[2] The origin of the dispute over 'climate justice' lies in this fundamental divergence from the Declaration of the Stockholm Conference on the Human Environment which included this principle as an article but since it was just a declaration, it has no legal repercussions. Climate justice has always been widely debated by the developed and developing countries and during one of the summits (at Copenhagen) one of India's senior negotiators argued, "Every human being has an equal right to resources of the atmosphere; therefore you have to take the per capita approach taking also into account the historical emissions."[3] On the second issue, while developed countries endorse sharing of the burden of reducing emissions, developing countries accuse developed countries of using climate change as a plain excuse to prevent the former from achieving the developmental levels of the latter after causing enough havoc through uninhibited emissions during the latter's developmental process. Though China is currently the largest emitter of GHGs, its per capita emissions are still far lower than most of the developed countries due to its high population. Therefore, questions over per capita emissions and population have always created further divisions between the industrialised and emerging countries. As Matthew Paterson and Michael Grubb observe, "The North-South divide reflects fundamental differences in priorities, responsibilities and perceptions."[4]

Based on the theory of EKC, developing countries have been arguing that many environmental problems are caused and aggravated during the industrialisation of a country; however, once the country attains a certain level of income, the trend often overturns, producing a curve shaped like an upside-down U; thus, developing countries must be first allowed to reach the level of development corresponding to that of developed countries and then stabilise their emissions and eventually reduce the emissions drastically. Based on this theory, developed countries too have justified their historical emissions and argued how they are on the path towards environmental sustainability. Each bloc's stand was rejected by the other. The EKC is a deviation from the old notion that economic growth endangered the planet. The concept was an offshoot of Grossman and Krueger's study of the potential environmental impacts of NAFTA (SO_2, dark matter and suspended particles), as well as Shafik and Bandyopadhyay's background study for the 1992 World Development Report. The argument that 'greater economic activity inevitably hurts the environment is based on static assumptions about technology, tastes and environmental investments' and that 'as incomes rise, the demand for improvements in environmental quality will increase, as will the resources available for investment' has been clearly underlined in this Report. The promulgators of the EKC have also maintained that the decline in environmental degradation would be triggered by 'structural change towards information-intensive industries and services, coupled with increased environmental awareness, enforcement of environmental regulations, better technology and higher environmental expenditures'.[5]

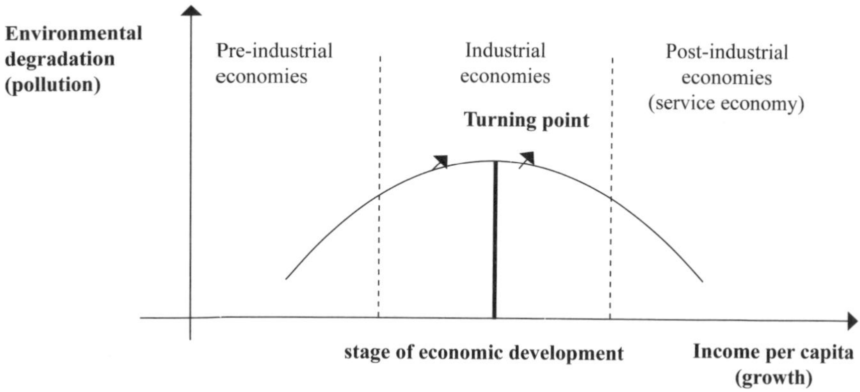

Source: Panayotou (1993)

Source: "Economic Growth and Environmental Quality: An Econometric and Decomposition Analysis," *Management of Environmental Quality: An International Journal* (Emerald Group Publishing Limited: 2007).

However, the EKC accounts for pollutants such as sulpher dioxide, nitrogen oxide, lead, DDT, CFCs and other chemicals released directly into the air or water. The reason why the EKC does not apply to carbon emissions is because all the other pollutants are known to have a more localised impact, whereas CO_2 emissions are known to have a global impact. Therefore, countries are forced to react at the local level while they pollute without impunity at the global level. This is the reason why rich industrialised nations, where only 25 per cent of the world's population lives, emit more than 70 per cent of the total global CO_2 emissions and consume 75-80 per cent of the resources of the world. Moreover, developed countries' carbon emissions have not decreased in accordance with the EKC.[6] There also has been a debate whether the EKC could be applied in the cases of both the developed and developing countries as their priorities and properties are diametrically opposite to each other at a particular point in time. In a capitalistic system, labour and natural resources of developing countries are exploited by the human capital and manufacturing capital-intensive industries of developed countries, which shift environmental degradation from the latter to the former; this is the cause of reduction in environmental degradation in developed countries. This criticism is closely linked to the laws of specialisation and international free trade. The biggest criticism against the EKC has been that it completely negates the possibility of 'environmental change negatively impacting economic activity in the early stages of development.[7]

However, the most pressing issue between the two groups of countries has been financial resources and technology transfer. Developed countries argue that developing countries should not commit the same mistakes that they did while industrialising themselves. This argument stands ground only if developed countries agree to share their knowledge resource with developing countries since they have more financial and technological resources and know-how than the latter. It is true that developing countries are obliged only to 'establish national inventories of their GHG emissions sources and sinks'; at the same time, one also has to keep in mind that they have 'little incentive from the convention to limit their emissions growth'.[8] One of the mediums of financial transfer, the GEF itself has been surrounded by controversy since its inception as developing countries object to the spending being dictated by the voting procedures of the World Bank and not the Conference of Parties. While developing countries have overtly termed climate change negotiations as an 'economic conference' instead of an 'environment conference', developed countries, particularly the US, have time and again insisted on using the term 'technology cooperation' instead of 'technology transfer'.[9]

I will do it my way!

The second conflict could be characterised by the differences of opinion between industrialised countries. Countries such as the US, Canada, Australia, New Zealand, Iceland, the Russian Federation, Ukraine, Norway and Japan (also called the Umbrella Group) have had differences with their European counterparts about the emissions reduction strategies. While Japan supports the sectoral approach (bottom-up) to tackle climate change, the US espouses another bottom-up approach by proposing National Strategies instead of the 'targets and timetables' approach backed by the EU. The sectoral strategy was quashed by a majority of developing countries and some of the developed countries due to uncertainties regarding choice of sectors and technology transfer in each sector. This approach focuses on promoting optimal energy use within each industrial sector by establishing multiple sector-specific negotiations that could also prove to be cost-ineffective. The classification would broadly be energy-intensive industries, domestic sectors such as electricity and transport, and household and commercial sectors.[10] The US, on the other hand, devised an approach by which countries pledge to undertake certain actions with compliance being assessed domestically.

Some experts contend that 'a strong top-down approach would involve strong global coordination, be centred around the pursuit of a common objective and be implemented through targets and timetables based on commonly agreed rules which would be progressively broadened and strengthened over time and would be legally binding with a strong measuring, reporting and verification (MRV) system and compliance mechanism'. Conversely, 'a weak bottom-up approach would have little or no global coordination of efforts, with coordination confined to, perhaps, a small group of countries'.[11] By raising the issues of urgency of the problem, transparency at the international level and cost-effectiveness, this group has discredited the arguments of those is in favour of the bottom-up approach. The latter believes that any action that could be implemented at the 'local, city, regional or single-nation level'[12] need not be pegged to any global regime. The top-down approach could not only take away the attention of the policy-makers from adaptation, which is a grave concern for the majority of countries in the world currently, it could also lead to compromises and the result would always be the lowest common denominator. Interestingly, there are contrarian views that neutralise the top-down versus bottom-up debate. Matthew Paterson remarks, "I am firmly opposed to thinking of it as 'top down' versus 'bottom up'. This is very unhelpful language. It was developed by Americans wanting to reposition the US for a post-Bush era, but realising they needed to paint Kyoto in negative terms, and what they came up with was the 'top down' or 'centralised' language. But it makes no sense in my view. Kyoto was very much 'bottom up'—a set of agreements by states—not a 'top down' imposition. The question is still what sorts of commitments will states make and how do they design an agreement that (a) builds trust that all will keep their promises, and (b) if we're lucky, creates a whole that is more than the sum of the parts in terms of emissions reductions – that is, enables states to be more ambitious than they might otherwise be."[13]

Who has more energy and who is more vulnerable?

Some of the other fault-lines which have dominated the discourse on climate change negotiations are worth mentioning. One among them is the split between the major energy producers and others. Countries which are dependent on energy exports (OPEC countries) or have developed a strong 'energy culture' (the US and China) over the years fear the negative consequences of emissions cuts on their economies. The latter group of countries has developed on account of large reserves of low-cost domestic fuel resources such as coal. On the other hand, countries that are dependent on imported energy such as the European countries are advocates of strong action on climate change as that would have direct positive repercussions on their balance of payments. The division between those countries which are more vulnerable to the direct outcomes of climate change such as sea-level rise or erratic rainfall patterns, and those which are relatively less vulnerable based on their degree of exposure to climate change as well as their economic resources to cope with the disasters—characterised by the labels of 'climate-ready' and 'climate-not-ready' respectively—is the next most prominent fault-line.

Countries such as Maldives and Tuvalu might be wiped off the map of the earth whereas the costs of adaptation to climate change in the US were considered insignificant for a long time and this perception has not changed drastically to this date. This is not the case in Europe, wherein the cognizance of the problem and the high costs of adaptation is reportedly higher than in the US. However, this perception in the US is expected to change. In the past one decade, the US has incurred losses worth more than US$ 257 billion due to hurricane damage only.[14] These losses may be much lower than the US' spending on the war on terrorism, yet the figures are significant since the country has already lost US$ 35 billion due to floods, tornadoes and heat waves in the first half of 2011 alone.[15] Countries such as Bangladesh and the Netherlands are equally vulnerable to the effects of climate change.

Yet, due to the huge difference in the resources these countries possess to cope with damage caused due to climate change, the Netherlands is in a better position than Bangladesh. For these reasons, Maldives, Tuvalu and Bangladesh have been more strident in their demand for a stronger global convention. Among the developing countries, the AOSIS and some of the LDCs would like to securitise the issue in order to fo[illegible] rest of the world to adopt immediate and substantial cuts in the emissions of GHGs instead of mere stabilisation targets; while Newly Industrialised Countries of East and Southeast Asia as well as emerging economies such as India and China would like to restrict the ambit of discussion to development and access to technology.[16] The securitisation debate has been mostly driven by the developed countries with firm backing of the AOSIS and the LDCs even in the UN Security Council (UNSC).

Why politics and science do not complement each other?

Besides these differences between nation states, there have always been differences of opinion between policy makers and scientists as the corollaries are still not completely traceable or predictable. A significant minority still questions the credibility of the claims of the scientific community – that of the IPCC, especially in the wake of the inaccurate claim made by it that the Himalayan glaciers could melt away by 2035. The Climatic Research Unit email controversy or more popularly known as the 'Climategate' revealed that scientists manipulated data in order to silence sceptics. This also sparked the debate just before the Copenhagen Summit whether climate change is a scientific, political or economic conspiracy to fulfil a particular section's interests. There have also been instances of differences with regard to the degree of impacts of climate change between the two communities. Vladimir Putin, the Prime Minister of Russia, opened a can of worms by commenting, "An increase of two or three degrees wouldn't be so bad for a northern country like Russia. We could spend less on fur coats, and the grain harvest would go up."[17] There are contrary views with respect to climate scepticism within climate negotiations. According to journalists such as Nitin Sethi (Times of

India), one of the major causes of the recurrent debacles at the negotiations is climate scepticism,[18] while experts such as Dr. Lavanya Rajamani insist that the international community has come a long way from climate scepticism and is definitely moving towards striking an international agreement.[19]

India's role in the creation of the global climate regime

Even before the negotiations started in 1992, the former Prime Minister, Rajiv Gandhi ha[illegible]ce tried to galvanise the support of the South to block international schema guided by the West to protect the environment. He argued that unless and until developing countries are provided environmentally friendly technologies at affordable costs, they should not stop growing economically. Interestingly, Dr. Rajendra Pachauri presented one of Rajiv Gandhi's ideas—that of a Planet Protection Fund (PPF)—at an international meeting to facilitate this.[20] Similarly, India was one of the first countries that recommended the international community to bring the GEF under the Conference of Parties (COP). Another Indian proposal was to introduce the concept of per capita equity by which 'an equitable solution' can be found on the basis of 'significant reductions in levels of per capita emissions in developed countries, so that over a period of years these converge with rising per capita emissions in developing countries'[21] as enunciated by Chandrashekhar Dasgupta, one of the most seasoned climate negotiators for India for nearly 15 years. The efforts made by India to lead the developing world and acquire a favourable framework convention before the Earth Summit in 1992 were thwarted by the rest of the developing countries including China.

A lot of ideas put forth by India were struck down by the developed countries, and even other developing countries, due to a lack of consensus among the latter. Developed countries had no intentions to assist the developing countries even long before countries such as China and India reached a position in which they could offer some amount of competition to the former. However, concerted efforts on the part of the latter to put pressure on the former would have produced some positive results for the latter. Instead, the developing countries got divided among themselves on the basis of their interests, whether it was in the form of founding the AOSIS based on the impact of sea level rise on small island states, or forming the BASIC on the basis of the developmental agenda of the countries that constitute it. The AOSIS came together primarily because their interests lie in addressing the effects of climate change and not the costs of strategies, which were of a bigger concern for India and China.[22] Therefore, they supported the idea of international negotiations through which they would receive more visibility; it would also provide them a platform to cooperate with industrialised countries.

The terminologies of Navroz K. Dubash—'Growth First Stonewallers', 'Progressive Realists', and 'Progressive Internationalists'–are fundamentally driven by their mottos—'It's our turn'! 'It's an unfair world'! and 'Seize the moment'! respectively.[23] These mottos are the propellers of India's strategy at the climate change negotiations. On the grounds of sovereignty, equity and development, India has asserted its position that it would not take on the burden of international obligations and that the developed countries should be ready to provide financial and technological assistance to the developing countries. India has also strongly criticised the domination of bodies such as the IPCC by the developed countries;[24] the inability of the developing countries to participate in the functioning of the IPCC in a holistic manner arose due to paucity of resources of all types—technological, financial, institutional and human—that are the prerequisites for participation in the IPCC. Ironically, it is an Indian (Dr. Pachauri) who leads the IPCC team presently.

The Chronicles of the Kyoto Protocol and the Bali Action Plan

As the Journey Reached Kyoto

The Earth Summit was followed by COP-1 in Berlin in 1995. The 'Berlin mandate' recognised the necessity for the implementation of stronger commitments through a feasible protocol. COP-2 held in Geneva in 1996 that called for legally binding mid-term targets generated momentum for the enforcement of a protocol, which materialised at COP-3 held in Kyoto, Japan, in the form of the Kyoto Protocol after ten days of rigorous negotiations. This imposed legally binding overall reduction in GHG emissions of 5.2 per cent below the 1990 levels throughout the period 2008-12. Some of the Annex I countries that were assigned individual targets include Canada, Hungary, Japan, Poland (6 per cent), the EU, Switzerland, most Central/East European states (8 per cent), and most importantly, the US (7 per cent). Interestingly, some of the other developed countries could increase their GHG emissions, for instance, Iceland (10 per cent), Ireland (13 per cent), Portugal (27 per cent), Greece (25 per cent), Spain (15 per cent) and Australia (8 per cent).[25] The Kyoto Protocol also paved the way for an international emissions trading regime that allowed industrialised countries to buy and sell emission credits amongst themselves; the Joint Implementation (JI) procedure enabling industrialised countries to implement projects that reduce emissions or remove carbon in another Annex I country in exchange for emission reduction credits; the CDM permitting developed countries to finance emissions-reduction projects in developing countries and receive credit for doing so.[26] The Kyoto Protocol, therefore, fits perfectly into the top-down model according to the pro-top-down approach advocates.

A total of 178 countries reached a binding agreement, though the US has still not ratified the protocol despite agreeing to a 7 per cent reduction target at Kyoto after it successfully bargained for a tradeable permit system. Neither Bill Clinton nor George W. Bush sent the protocol to the Congress for ratification, with the latter openly rejecting it. Neither has Barack Obama, despite being more responsive to climate change concerns, taken any step in this direction since becoming the 44th President of the US. It was on July 25, 1997 that the US Senate unanimously passed the Senate Resolution 98 (also called Hagel-Byrd Resolution), quashing any possibility of ratifying any treaty that would (a) impose mandatory GHG emissions reductions for the US without also imposing such reductions for developing nations, or (b) result in serious harm to our economy.[27] This was a huge setback for climate change negotiations, particularly to the implementation of the Kyoto Protocol that eventually came into force on February 16, 2005 without the US' stamp on it. As Christoph Böhringer and Carsten Vogt find out in their 'back-of-the-envelope analysis', "The recent changes to the Kyoto Protocol —most notably, US' withdrawal and unrestricted emissions trading—boil down climate policy to business as usual without any compliance costs for participating countries."[28] It is true that without the US' ratification of the Kyoto Protocol, though the first commitment period target could be achieved but it would set a dangerous precedent to rest of the industrialised countries which would like to withdraw from it too. Japan has already voiced its displeasure with the Protocol. It is also impossible to address an issue like climate change without the principal contributor being brought to book for both its historical as well as contemporary responsibility for the problem. It was the US' diplomatic manoeuvring that resulted in the unanimous endorsement of emissions trading and sinks within the purview of the Protocol, yet it washed its hands off its commitments. Emissions' trading has been called by environmentalists as a license to emit or pollute. The Kyoto targets would be met by the majority of countries due to these flexible mechanisms and no 'real' reduction in the emissions as envisaged by the UNFCCC would take place. At the end of the day, the aggregate emissions would not have reduced a great deal from the business as usual. However, one has to remember that the world is not idealistic and unless countries or industries are rewarded with incentives to reduce emissions, no one would be ready to sacrifice their interests. To complete the profit-and-loss cycle, the norm of paying the price to another country or industry for not being able to meet the requirements operates. There are other reasons why the Kyoto Protocol could not produce all the desired results. Some of them include:

- The parties did not address the issue of adaptation and invested all their energy in addressing mitigation as they felt the former would dilute the need for the latter. One could also treat this as a lack of scientific maturity as it has already been made clear without adaptation, mitigation would not be possible. The first priority always has to be to develop the coping mechanism so that the causes of the symptom could be counteracted.
- The attempts to find solutions to the issue by trying to fit it into other international models such as the disarmament regime have proved to be counter-productive. The debate always boils down to 'who will lead the rest of the world in making the right choices and undertaking the right actions'?

- Many allege that the biggest flaw in the Kyoto design is that it has neither elucidated the objectives nor the means to achieve these objectives.[29]
- The implementation of the Kyoto Protocol, to a large extent, is also considered to be contingent on the success of emissions trading and the ability of such a system to raise emission permits worth more than a trillion dollars.
- The means of monitoring and enforcement have not been given adequate attention in the document. Being a unanimous international agreement, these concerns are essentially negated.

Tryst with the Offsprings of the Kyoto Protocol

The Kyoto Protocol, while creating room for cooperation between the different blocs, left several questions unanswered and partially answered. CDM is a classic example of a well-intended step that has neither been utilised to the degree it should have been nor has it been employed in an equitable and professional manner in a majority of cases. The CDM is fraught by several other impediments besides the ones discussed in the Indian context in the second chapter.

- Developing countries have insisted that instead of having a unilateral CDM in which the pricing of technology is solely in the hands of the supplier, a compliance mechanism in the form of a technology acquisition fund should be in place to which the defaulting Annex I parties could contribute to facilitate technology and finance transfer to developing countries.[30] This idea has so far been not heeded by industrialised countries.
- The primary objective of the CDM is to synergise reduction of GHG emissions and contribution to sustainable development in the developing countries. Several developing countries have either not set the sustainable development criteria and standards due to a lack of institutional and technical frameworks, which makes them susceptible to exploitation by industrialised countries, while others have simply not mainstreamed climate change in their sustainable development policy as a result of which both introduction and implementation of the CDM projects become difficult apart from the fact that the role of private players becomes negligible.

- Among the technical impediments, the baseline setting, methodology and additionality are the most prominent ones. A significant proportion of the CDM methodologies put forth by the developing countries are rejected by the Executive Board on account of lack of approved methodologies in the sectors that interest the developing countries, which makes the task of setting baselines difficult. The additionality factor—that the project would not have occurred in the absence of the CDM—is difficult to prove in many projects.[31] Based on the technical difficulties in these two factors, concerns amongst developing countries in baseline setting have also blocked the expansion of the scope of the CDM from a purely project-based approach to sector-based or programme-based approach. India and China are among them though they ardently champion the idea of expansion.[32]
- As Shyam Saran points out, in countries where the distribution of targets for various industries has not yet taken place, and incentives for the reduction of emissions and project implementation are not given, all the benefits of the CDM cannot be reaped. When there are no incentives, the price of Certified Emission Reduction (CER) goes down naturally.[33]
- The future of the CDM itself is in jeopardy as its progenitor, the Kyoto Protocol is not currently popular among industrialised countries. This affects long-term projects particularly, such as forestry, in which the competition is already dismally low across the spectrum.

Another important aspect of the Kyoto Protocol is Transfer of Technology (ToT). Though technology transfer was not officially envisaged under the CDM, it could indirectly aid ToT through FDI. However, progress has been slow mostly due to the disinclination of private sector players of the developed countries to invest in developing countries due to the fear of losing 'international competitiveness'. The costs of these technologies are too high that do not get compensated by the revenue generated through CDM. A majority of developing countries including India have rebuffed the idea of diverting Official Development Assistance (ODA) for CDM or even the creation of a fund under ODA to promote CDM projects. Another glitch in the process of deployment of technology is the disputes regarding the Intellectual Property Rights (IPR).[34]

When one looks into the relationship of industrialised countries with the Kyoto Protocol, one comes to the conclusion that the European countries seem to be the only fervent supporters of the negotiations since their monopoly over green technology and the carbon market is unmatchable. The US has still not ratified the Kyoto Protocol though it is a signatory. Since it has only signed the treaty and not ratified it, it has only declared the intention of making the terms of the treaty legally binding on itself and not made the terms of the treaty legally binding. It is nowhere near fulfilling its first commitment period target. Japan has uncompromisingly stood by its position that it would withdraw from the historic Kyoto Protocol reached upon on their land, leaving a huge question mark over the second commitment period. This makes Canada the only G-8 country that would well fall short of fulfilling its target of the first commitment period of the Kyoto Protocol since the other industrialised countries have managed to pull up their socks with the assistance of various mechanisms introduced under the purview of the UNFCCC to meet theirs. Currently, apart from Greece and a few Economies in Transition (Central and Eastern European countries that are undergoing the process of transition to a market economy), Canada is the only country to be in violation of the compliance principles of the Kyoto Protocol. Australia, after prolonged criticism from both international as well as domestic constituencies ratified the Kyoto Protocol only in 2007. Thus, it is the US that has been the biggest irritant in the negotiations since it is one of the largest polluters, and is considered the predominant pole in a professed multi-polar world.

The Post-Kyoto Picture

Until 2001, the negotiations focussed mainly on the strategies of industrialised countries to reduce emissions. The Marrakesh Accords (2001) that came into force in 2005 were rejected by the US and even countries like Australia, which were rooting for carbon sinks in forests and agriculture as well as unrestrained emissions trading. In fact, in 2000, the split between the energy-rich and the imported energy-dependent countries had led to the suspension of talks. Not only did the Middle-Eastern/West Asian oil giants resist cuts, the parties also failed to reach consensus on the financial mechanism as well as consequences of non-compliance by countries. The differences between countries such as the US, Australia, Japan, Canada etc on the one side Europe on the other took the backseat thereafter. Post-Marrakesh talks have been largely dominated by the developed versus developing country positions debate. At Montreal (COP-11), the dialogue on the two-track approach for the future was initiated – first, to negotiate the second period commitments for the developed countries under the Kyoto Protocol implemented by the Ad Hoc Working Group-Kyoto Protocol (AWG-KP); second, to discuss the adoption of new commitments by the Annex I parties along with Nationally Appropriate Mitigation Actions (NAMAs) by the developing countries under the Convention implemented by the Ad Hoc Working Group-Long-Term Cooperative Action (AWG-LCA). The two-track process fructified at COP-13 after the dialogue came to an end. Held in 2007 at Bali (Indonesia), COP-13 laid down the Bali Action Plan for a post-2012 climate change regime, that is, the Convention track. For the first time after 2001, US became an integral part of the negotiating process without being a responsible partaker of the Kyoto track. The mandate of the AWG-LCA was also to devise MRV system for national mitigation actions and to fix the financial mechanisms for adaptation and technology transfer. This track was supposed to conclude in Copenhagen with an 'agreed outcome'; instead it could only generate an outcome, which could be dubbed a third kind of instrument – a high-level political instrument. The other two instruments are – first, a legally binding instrument (Example: Kyoto Protocol) and second, an instrument with limited legal status (Example: Berlin Mandate). The Marrakesh Accords are broadly the rules required to implement the decisions enshrined in the legally binding and non-legally binding instruments such as GHG mitigation targets, the CDM, compliance costs of the Kyoto Protocol, etc.

Copenhagen and Cancun: The Geopolitical Dead-end of Climate Change Debate?

The Ultimate Political Showdown at Copenhagen

COP-15 took place in Copenhagen in 2009, in which the negotiations embodied an international system that some would say was tilted equally to the developing countries and their demands. The purpose of the Summit was to agree upon a mechanism for the period from 2012 when the first commitment period under the Kyoto Protocol expires. This was certainly one of those occasions in which the differences between the developed and developing countries' positions became glaring. The voices of developing countries were heard as much as that of industrialised countries; this could well be the reason why instead of coming up with a post-Kyoto mechanism, it could only produce a 'political agreement' that had no legal subtexts. Moreover, the 13-paragraph Copenhagen Accord is considered an external document as it was not negotiated within the UNFCCC process. This could be viewed as an outright dilution of the Bali Action Plan that had called for the full implementation of the Convention. The Copenhagen Accord was drafted by the BASIC countries as well as the US that recognised climate change as one of the greatest challenges of the present day and recommended actions to keep temperature increase below 2°C. The Copenhagen Accord also contains provisions for market-based incentives for low carbon growth, adaptation funds for developing countries, the LDCs and the Small Island Developing States (SIDS), plus the REDD+ (Reducing Emissions from Deforestation and Forest Degradation) mechanism by avoiding deforestation and forest conservation.

Even before the Copenhagen Accord came to light, a leaked Danish text (also known as the Copenhagen Agreement) drafted by the US, the UK and Denmark raked up controversy as it spoke about abandoning the Kyoto Protocol. The Summit also witnessed an array of NGOs—environmental, socio-economic and security-related—from across the world trying to impress the world leaders. For example, an NGO named 'Chintan' took a group of rag pickers from India to Copenhagen to inform the leaders about the contribution of waste recycling to emissions reduction and the importance of rag pickers so that it would be incorporated in the CDM.[35] Similarly, a group of NGOs—Greenpeace, WWF, German Watch, David Suzuki and others—brought out an NGO treaty that could run parallel to the Kyoto Protocol depending on the allocation of carbon space and finance.[36] The role of NGOs would further be elaborated upon in a subsequent chapter.

Reading Between the Lines

The Copenhagen Accord confronted enough criticism for not being legally binding, for not setting real targets to achieve in emissions reductions and for the scepticism over the mobilisation of US$ 100 billion every year by 2020 by the developed countries. Some of them have been explained below:

- By stating that 'the funding will come from a wide variety of sources, public and private, bilateral and multilateral, including alternative sources of finance',[37] the Accord left a huge question mark over the contribution of each developed country and the credibility of the sources. The inclusion of developing countries in the governance structure of the Copenhagen Green Climate Fund (GCF) came as a welcome development, but the Fund has still not become functional. The creation of the Fund basically was an attempt on the part of industrialised countries to indulge in the geopolitics of divide and rule by which emerging economies were effectively demarcated from developing countries. With this, the doors of the possibility of any assistance for mitigation and adaptation for countries such as India, China and Brazil were closed.
- The peaking of emissions mentioned in the document has also been questioned by several scientists. Developed countries are expected to reduce their emissions by 80 per cent by 2050 as a part of worldwide goal of a 50 per cent reduction in emissions. Scientists have found that if the world's emissions do not peak by 2020, the international community would not be able to meet its 2050 target. Besides, it would also not be able to keep temperature rise below 2°C. Studies suggest that there is a 50 per cent chance of achieving this if the peaking happens in 2018 and if emissions shrink by 4 per cent per year thereafter and if the peaking happens two years later, the decline would have to be 5 per cent per year; beyond 2025, there is no chance of achieving this target.[38] Though developing countries agreed to discuss this issue, it is not clear as to how their emissions would peak and decline by 2050 at a juncture when they do not even have any obligations to reduce emissions.
- Another important development was the introduction of a domestic MRV for the developing countries in order to make them accountable

for their domestic mitigation actions by making the provision for 'international consultation and analysis'.[39] This has also been stonewalled by developing countries for a long time as according to them under the principles of equity and 'common but differentiated responsibility', all their actions would be voluntary and not subject to international scrutiny. However, one should admit that biennial national communications need not necessarily impinge on national sovereignty. National communications are nothing but reports on the steps countries take or intend to undertake to implement the Convention. The use of the terms, 'consultation' and 'analysis', need not necessarily imply 'verification' by any external agency. This provision is meant largely to address the issue of transparency on the part of developing countries at a time when the pressure on emerging economies to adopt emissions reductions is mounting. The opposite view maintains that MRV, one of the pillars of negotiation, is about equality despite disparity and not about transparency.[40] Phrases such as 'meaningful mitigation actions and transparency on implementation' could be exploited by the developed countries to coerce the developing countries to toe the line of the former on not only climate change issues but also other international issues. Also, in an asymmetric world in which some countries are more equal than others, the chances of imposing punitive measures against a country like the US are lower than against a country like Bangladesh. Therefore, international intervention, in any form, is nothing but fatal to the interests of developing countries.

- One of the biggest drawbacks of the outcome was that the parties hardly reached a consensus with respect to future emissions reductions by developed countries—whether it is the second period commitments under the Kyoto Protocol or the post-Kyoto commitments.

These are the reasons why the Copenhagen Accord has been branded a 'political agreement', with very less clarity on most of the crucial issues and more focus on the phraseology of the statements. The endeavour was more to reach a 'politically correct' and 'balanced' agreement and not a 'concrete' and 'futuristic' one.

The politics of adaptation has further aggravated the complexities of climate change negotiations with very few convergences between different countries belonging to different blocs. Adaptation has been described by many as 'the poor and derided cousin of emissions reduction'.[41] It was only in the 21st century and particularly during the second half of the first decade that adaptation found its rightful place on the negotiations table. For the first time, some of the most vulnerable developing countries let go off their suspicion that this is a clandestine attempt by the developed countries to prevent their growth. More importantly, the gap between the developing and developed countries in terms of their concerns regarding adaptation narrowed to a certain extent. Developed countries have stressed mostly on mitigation while developing countries have bestowed equal or more emphasis on adaptation. The divergence starts at the same question that has obstructed the progress of talks—who pays? It is interesting to note that there is great disparity in terms of the funds allocated for mitigation projects vis-a-vis adaptation projects under the GCF. For example, reports have disclosed that industrialised countries have apportioned far more fast-start finance funds for mitigation than adaptation; [42] these funds are meant for the LDCs and the SIDS.

Reactions and Shift in Relations

Several delegations including that of Tuvalu, Venezuela (who were represented by their respective Heads of State) and Sudan (Chair of G-77) threatened to walk out if the agreement did not mention legally binding emission cuts and temperature rise to well below 1.5°C. The summit was in reality reduced to a '25-member mini summit' with around 100 world leaders mostly belonging to African, Latin American and Asian under-developed and developing countries sitting in the plenary chamber next door believing that 25 rational parties would produce some sort of an encompassing document—plainly highlighting the almost trifling role played by those countries in the negotiations whose stakes are the highest in the evolving environmental and geopolitical scenario. Lumumba Stanislaus Di-Aping, the Sudanese negotiator said, "This is asking Africa to sign a suicide pact, an incineration pact in order to maintain the economic dependence of a few countries. It's a solution based on values that funnelled six million people in Europe into furnaces."[43]

The failure of the Copenhagen Summit was largely attributed to the global recession and conservative domestic pressure in the US and China. The redundancy of the Kyoto Protocol took the centre stage of discussions as the developed countries no longer wished to acknowledge the legal distinction between the developed and developing countries, especially at a time when the developing countries were growing at a fast pace, largely unaffected by the economic recession of 2008, while the developed countries were reeling under the hardships caused by it. In such a situation, in which China and even India have emerged as competitors economically, the possibility of a partnership between them was grim, especially since the developed countries would not be ready to provide any form of financial assistance to India and China. Industrialised countries did not approve of the two-track process and instead stressed on the need for a single binding agreement that would replace Kyoto under which both the developed and developing countries would have to share the responsibility for the emissions cuts. Developing countries still believed in the developed countries' supreme responsibility for both causing climate change and tackling it. This Accord has been a major setback for not only climate change mitigation mechanisms but also for the UN whose role in the international climate deals was questioned for the first time.

China, which has been targeted for being the spoilsport in the COP-15, was ironically the first country in the world to ratify the UNFCCC in 1992. However, China is one of those countries that joined the Asia-Pacific Partnership on Clean Development and Climate, an anti-Kyoto mechanism led by the US. However, it went on to ratify it in 2002. The other countries that have joined this body are Australia, Canada, Japan, Korea and not surprisingly India. This is a purely industry-oriented mechanism that only aims to improve relative environmental cleanliness of the GHG-emitting industrial and energy development sectors. China's international environment policy has been formulated on two grounds—firstly, non-compromise on economic development and secondly, avoidance of high-cost commitments.[44] China, like the US, believes in the bottom-up approach.

China, for example, unravelled its first action plan on climate change in 2007 'to reduce energy consumption per unit of gross domestic product by 20 per cent by 2010 and to increase the share of renewable energy to some 10 per cent, as well as to cover roughly 20 per cent of the nation's land with forest'.[45] China's role in the climate change negotiations is paramount as the US and China account for over one-third of the GHG emissions of the world. With the Copenhagen Summit, the concept of G-2 evolved, placing the US and China at the forefront of global decision-making. One could say that this is a sign of the emergence of a truly multi-polar world as China is exceedingly making its presence felt all across the globe not just economically but also politically. The other stakeholders are Europe, Russia and India. In climate negotiations, the AOSIS and G-77 are equally important players as their pressure drives industrialised countries as well as emerging economies into taking at least the lowest common denominator. The European media openly criticised Wen Jiabao for stalling the talks to protect China's double-digit growth figures, and Barack Obama for stabbing the EU in the back by letting China go without making any legally binding agreements, respectively especially since the EU had pledged to reduce GHG emissions by 80 per cent by 2050.

The French President, Nicolas Sarkozy and the German Chancellor, Angela Merkel were particularly disturbed by the show of power by India and China as well as by Obama's diplomatic manoeuvring, which in a way, sidelined the European powers that had played the most important role in the negotiations since they started. A section of strategic thinkers saw the developments at the Copenhagen Summit as something which visibly showcased the emergence of a new political order, 'one shaped by the new self-confidence of the Asians and the powerlessness of the West'.[46] India sent out a strong message to the West that it could no longer be dictated by the financial mechanisms designed by the West aimed at granting the responsibility of the administration of the Green Fund to institutions like the World Bank or the IMF, which continue to be governed by the financial contribution of each country and thus biased towards the West for obvious reasons. It also raised the issue of 'climate justice' in response to the demands of the developed countries to make the developing countries' domestic commitments legally-binding and verifiable. China, on the other hand had multiple agendas—to engage in the 'climate market' negotiations as it is regarded as the biggest beneficiary of the CDM, as well as to reiterate its indisputable role of leadership in the developing world. Both the Indian and the Chinese contingents also made their conditional association with the Accord clear. They asserted that the primacy of the UN process centred on the two-track dialogue should not be bypassed in order to include a third track based on the Copenhagen Accord that had not been endorsed by a majority of countries.

At the outset, one could haul out two major shifts in geopolitics and international relations—the ascendency of the BASIC and the marginalisation of the West, especially Europe. The BASIC countries exhibited their disapproval a number of times during the Summit due to the refusal of industrialised countries to dedicate enough attention to the second commitment period of the Kyoto Protocol. They also broke ranks with the G-77 in order to broker an agreement. In effect, these countries demonstrated that since they are making a transition from a developing to a developed country, they are required to be in both camps to strike a compromise formula. The G-77 itself got divided into various fragments as the Green Fund clearly draws a distinction between the emerging economies and the LDCs and the SIDS. At Copenhagen, the US had hardly any role to play initially as Shyam Saran says. With mere voluntary targets of the scale of 17 per cent emissions cuts by 2020 that are not subject to international compliance despite huge technological, technical and financial resources, the US has indeed set a very low standard despite continuing to be one of the largest emitters in terms of both aggregate and per capita emissions. Ideally, the EU and the developing countries irrespective of their levels of development should have joined hands, but the EU squandered the golden opportunity because of its response to emerging economies' reservations about the MRV and possible emissions cuts.[47] The rigidity of the EU was counteracted by the flexibility of the BASIC, which made a few compromises that have already been reflected upon earlier in the chapter. The leadership of the BASIC was the biggest highlight of the Summit. While talking about cooperation between the BASIC countries, one should also keep in mind that China is far ahead of the rest, especially India, in terms of 'total GDP, annual rate of growth in GDP, carbon intensity of economy, aggregate and per capita emissions, rate of growth in emissions and the contribution its emissions make towards the current rate of build-up of CO_2 in the atmosphere'.[48] Therefore, close engagement with China on this issue could prove to be counterproductive for India.

There was no significant change in India's position and perception with respect to climate change until the Copenhagen Summit. At Copenhagen, India had to modify its stand and state its global climate change policy according to the changing geopolitical scenario. Shyam Saran, who was the Prime Minister's Envoy of Climate Change at the summit, outlines India's red lines when it comes to its stand on climate change at the negotiations level. First, the global climate regime should not be constrained and should also take into consideration growth prospects, eradication of poverty, current reliance on fossil fuels, the costs involved in the transition to cleaner fuels and the need for technology transfer and financing. Second, mitigation of climate change has to be carried out in India's interest but it is contingent largely on a supportive global regime. Therefore, it is in India's interest to negotiate. Third, the principle of equity should be supplemented by three mandates—implementation of voluntary measures developing countries to reduce carbon intensity of growth and absolute reductions by the developed countries; mainstreaming of poverty eradication; and a two-track dialogue (Long-Term Cooperative Action and the Kyoto Protocol) so that the Kyoto Protocol is not be abandoned. Fourth, the question of where is the money going to come from has to be addressed with seriousness. India had to opt out of being one of the major recipients of fast-track finances so that the LDCs and the SIDS would get the requisite resources for both adaptation and mitigation. Fifth, the per capita principle should be applied by which India would not have to adopt any legal obligation. India is ready to take concrete actions and is ready to register them with the UNFCCC. International MRV for internationally aided actions is acceptable since India's actions are transparent. Last, India should focus on adaptation more because the impacts of climate change would not be curtailed even if emissions become nil owing to the accumulation impact. Development would contribute to the capacity to adapt.[49]

As the Ice Floated to Cancun

After a much hyped Copenhagen Summit that produced hardly any tangible results, the COP-16 at Cancun was deemed a success by most of the analysts. A lot of steps envisioned in the Copenhagen Accord were reformed and formalised through an official UN Agreement. The two-track process was brought back into focus while keeping the decisions arrived at Copenhagen at the forefront of negotiations. The international community took some unprecedented decisions and buried the failures of Copenhagen. At the same time, the Accord has its shortcomings. Many analysts consider it as yet another lowest common denominator that showcases compromises and confusions regarding the nitty-gritty and implementation; while others perceive it as a good foundation for moving forward.

- The international community pledged to keep the global temperature rise below 2°C. At the same time, it also agreed to reconsider whether the figure should be reduced to 1.5°C in 2015—a long-standing demand of the LDCs and the SIDS.
- Both the developed and developing countries reached the consensus that the global emissions of GHGs have to peak 'as soon as possible'[50] and that despite the fact that the developed countries would take the lead in terms of cutting carbon emissions, developing countries would also be required to reduce their emissions. The Agreement provides emissions cuts targets for approximately 80 countries including China and India. This could clearly be earmarked as a major shift in stand for a country such as India.
- The Cancun Adaptation Framework was established in order to enhance action on adaptation. The inclusion of 'common but differentiated responsibility' in this context makes sure that the onus lies on developed countries to fulfil their obligations.
- Interestingly, the 'international consultation and analysis' specified in the Copenhagen Accord were also formalised by making the mitigation actions of developing countries [also called Nationally Appropriate Mitigation Actions (NAMAs)], both 'domestically supported' as well as 'internationally supported', subject to a domestic MRV system; however, a new clause was added by which the 'internationally supported' actions

would be subject to an international MRV too. The question that would perturb the developing countries would be as to which body would be responsible for the international MRV. The international scrutiny could even be misused by the developed countries to take unilateral trade actions against developing countries, such as taxing their export of goods and services.

- REDD+ was finally given formal backing by which rich nations could pay poor countries for not cutting trees. It is not clear as to when the scheme would take off. Secondly, this could become yet another carbon offsetting market, making climate change negotiations more and more a game of hardcore economics.
- The developed countries pledge to contribute equally to funding adaptation and mitigation, with funding for adaptation prioritised for the most vulnerable developing countries, such as the LDCs, SIDS and Africa. As already proven by reports, this has not happened and the allocation of funds has been disproportionately in favour of mitigation. Moreover, the climate aid of US$ 100 billion and fast-track financing amounting to US$ 30 billion are not under the UN umbrella. The GCF would be governed by a Board of 24 members – equal representation from developed and developing countries. Besides, the Transitional Committee that will design the GCF would be constituted by 25 members from the developing country Parties and only 15 members from the developed country Parties. The decision to appoint the World Bank as the interim trustee of the Fund has not gone down well with the developing countries for various reasons that have been highlighted earlier in the chapter. The question remains as to how long this interim period could last.
- The Parties decided to establish a Technology Mechanism to facilitate both technology development and transfer. Under this, a Technology Executive Committee and a Climate Technology Centre and Network would be set up. The failure of the Parties to elucidate the source of finance or their location or even the time frame of its establishment has yet again left a host of questions unanswered.

The Parties, however, postponed the decision on the future of the Kyoto Protocol and its second commitment period to the Durban Summit to be held in 2011. The lack of enthusiasm among some of the developed countries such as the US, Japan and Canada with respect to the Kyoto Protocol is the primary reason for the deferment. Thus, for the first time since the beginning of the negotiations, the developed countries were kept at par with emerging economies such as India and China irrespective of their responsibility for the problem. There was also a general acceptance among the international community that some of the important initiatives to address climate change were expected to take shape outside the domain of the UN process. The efforts of countries such as Bolivia to keep economics or rather 'global capitalism' away from the negotiations were thwarted.[51] In fact, the Cancun agreement was also a result of manipulations. Initially, the Bolivian delegation raised objections against the attempt to convert the mock or informal plenary (consisting of observers and NGOs besides the country delegations) into the actual plenary (consisting of only country delegations) by the Mexican Foreign Affairs Minister, Patricia Espinosa on the grounds that it was not debated on. However, she went ahead and adopted the Agreements despite lack of 'consensus', if the true meaning of the word is taken into consideration. Therefore, consensus no longer implies unanimity, if a vast majority agrees. The question is whether the same principles would apply if a stronger country such as Japan or Australia objects to any agreement in the future. Such an exclusivist approach does not augur well for Durban (COP-17), especially since the document produced at Cancun is common for both Annex I and non-Annex I countries, which is a first in any case, and it does not mention legally binding emissions cuts.

Yet another Game of World Politics

At the Cancun Summit, the former Minister of Environment and Forests, Jairam Ramesh, raised a few eyebrows when he said in his speech at the high level segment that 'all countries must take on binding commitments in an appropriate legal forum', which was contrary to India's earlier position. However, later he clarified that his attempt was 'to walk the thin line between safeguarding our position while showing a level of sensitivity to the view shared by the majority of countries at Cancun, including many of our developing country partners'. He also made a distinction between an 'appropriate legal forum' and a 'legally binding agreement'.[52] The BASIC partners, Brazil and South Africa, switched their sides to the bloc of the AOSIS, LDCs, Africa and four of the SAARC partners (Bangladesh, Maldives, Nepal and Bhutan) that insisted on a legally-binding agreement. At the same time, the Indian establishment claims that it has safeguarded its interests by successfully convincing the other parties to delete the clause which wanted global GHG emissions to be reduced by half by 2050, which would have put the burden of fixed emissions cuts on developing countries like India as well. This would have curtailed India's right to equitable sustainable development. A document prepared by the Indian Government allegedly reveals that its push for easing IPR control on green technology was ignored. Somewhere down the line, the Indian Government feels that turning the Cancun Accords into firm actions by the end of 2011 might hurt the interests of the country. Earlier, Prime Minister Manmohan Singh had formally signed a declaration at the Major Economies Forum in 2009 along with other G-8 countries, which purportedly violated India's earlier position on 'equal per capita rights for all countries to the global common carbon space in the atmosphere'.[53] Therefore, India is said to have made a series of compromises in order to be part of the 'coalition of the willing', which was essentially the handiwork of the US meant to rope in as many countries as possible on its side to endanger the future of the Kyoto Protocol.

Clearly, there was a shift in the geopolitical scenario that was reflected at the Cancun Summit. The US and the EU, along with Japan, Australia and Canada had managed to keep the Kyoto Protocol hanging in the balance, while persuading the rest of the world to sign an agreement after the previous year's debacle. The marginalisation of the BASIC was imminent since the LDCs and the SIDS (including the AOSIS) vigorously demanded a formidable agreement that entails actions from both developed countries and emerging economies. Countries like Brazil have always had a flexible stand on the issue of legally binding emissions reductions as long as no serious constraints are imposed on them. South Africa decided to go with the rest of the African continent that called for stronger commitments from the entire world in the wake of recurrent droughts. China has embarked on the path of green growth in a full-fledged manner and it is one of the major hubs of the global carbon market. India was perhaps left with no choice but to make heavy concessions on its earlier stand so that it is not isolated on an issue that is of concern to the whole globe. It is also guided by domestic factors which in turn are driven by internal contradictions. India's quality of environmental governance is yet to reach an appreciable state, and considering that it is next to impossible that its emissions could peak before 2020 due to increasing energy demands, India was caught in a sticky situation.

As far as the LDCs and the SIDS are concerned, they are exceedingly dependent on other countries—both developed and developing—for their adaptation and mitigation measures. These countries are least responsible for the emission of GHGs, yet they would eventually be the hardest-hit victims of global warming. For example, the Maldivian President, Mohamed Nasheed, gave a rousing speech at COP-15, Copenhagen in which he reiterated, "Carbon concentrations higher than 350 parts per million, and temperature rises above 1.5°C, will submerge my country, dissolve our coral reefs, turn our oceans to acid and destabilise the planet's climate." He continued, "For global emissions to peak by 2015 as science demands, industrialised countries must raise their level of ambition. They must commit to collective reductions of 40 per cent by 2020, and 95 per cent by 2050. But developing countries must also do their bit. The rich world may have caused the climate crisis, by filling our atmosphere with pollution. But two wrongs don't make a right. Developed countries created the climate crisis; developing countries must not turn it into a calamity." In fact, he urged the developing countries, specifically China, to adopt quantifiable and international verifiable actions, and to reduce their emissions 30 per cent below business as usual by 2020.[54] The Male' Declaration on Human Dimension of Global Climate Change, presented at the international gathering of the SIDS in November, 2007 drew clear linkages between climate change and human rights; this reinforces the point that climate change deprives human beings of their right to live. This is the reason why the climate change policies of the AOSIS countries are centred on survival and sustainability. The Maldives was also caught up in an unfortunate controversy when the documents leaked by Wikileaks allegedly insinuated that the Maldives had agreed to back the Copenhagen agreement if the US were ready to provide it with US$ 50 million for adaptation projects.[55] This shows that the SIDS and the LDCs could easily be arm-twisted or bought by the rich countries. They could also be playing for politically and economically motivated 'national interests' as their chief demand is money. It was at the Cancun Summit that these countries played the be-all-and-end-all role in ensuring that the BASIC did not prevail upon the rest of the international community. By creating fissures within the BASIC and G-77, they also ensured that the roles of EU and the US got enhanced.

When the Reality Is Merged with the Political Economy

The climate change negotiations, some would say, is nothing but an extension of the struggle between the industrialised and developing countries to maintain their supremacy and to seize the reins of power as well as influence respectively. It is universally accepted that industrialised countries are responsible for global warming and climate change; therefore, they have to take the lead irrespective of the fact that the emissions of developing countries could exceed that of the developed countries in the current scenario. The negotiations, in a way, present the inherent contradictions of the climate change debate. While the issue is more environmental and socio-economic by nature, the negotiations have turned it into a complex game of unscrupulous politics and inequitable economics. This is the reason why many prefer to call them 'trade negotiations'. The vision of the Rio Declaration was to lead the whole world along the path of sustainable development. The international community has bypassed Agenda 21 and encased the debate within the coffin of temperature rise and carbon emissions, neglecting myriad factors that as as important in the climate change debate. Since the struggle for supremacy has invaded the negotiations too, every Summit along with the respective Agreement has become a relic of absurdity and short-sightedness. The art of negotiations and diplomacy has taken precedence over the real issues that are confronting humankind. This is reflected in the attitudes of the international community that has combined environment and diplomacy to transfer the decision-making powers to the diplomatic community and not to those who devise environmental policies.

The global climate change regime is divided along various lines despite agreeing on issues that require global attention. From the outside, these differences are based on mere technicalities with respect to strategies, while the objectives and ambitions are the same across the board. As one reads between the lines, it is not only about 'who pays', it is more fundamental; it is also about 'who will take the lead' in such a competitive world. The Kyoto Protocol paved the way for the developed countries to execute this onerous task but the mood has changed since then. Countries have taken a few initial steps to tackle the issue but there is a long way to go as these steps are not enough to prevent large-scale environment change, which is a bigger concern than climate change. Hardcore economics guides the climate change debate as country after country tries to reap profits out of the global carbon market. This would be a highly cynical view of the negotiations. While one section would like to play the wait-and-watch game, the other would like to securitise the debate in order to take bigger leaps in terms of resolving the issue.

It would be contentious to look at the debate from the perspective of a developed versus developing world as the negotiations have reached a stage in which the differences within the two blocs have also gained salience. The contradictions and disparities in the implementation of various tenets of the Kyoto Protocol have left a huge vacuum in the global climate change regime. The CDM is touted as a successful venture but the question of emerging economies would always be an encumbrance on future regimes. The uncertainty surrounding the LDCs, the AOSIS and so on has been sorted out to some extent by the decision of raising funds. This has also become a bone of contention with allegations of non-compliance and deviation of pre-existing foreign aid towards adaptation and mitigation funds becoming rampant. To add insult to injury, the theme of the future of the Kyoto Protocol was untouched both at Copenhagen and Cancun.

India's core perceptions and positions have been steadfast since the beginning. Due to international and domestic pressures, it has moulded them with the changing international system. As Navroz K. Dubash has rightly put it, India has definitely moved far from the growth-first stonewallers' position, which regards climate change negotiations as a geopolitical threat to its growth and phenomenon of climate change itself as a hoax. Currently, India's policy is dangling between the 'progressive realists' and the 'progressive internationalist' with equity still occupying the driver's seat.[56] The so-called bottom-up approach has been one of the pillars of India's climate change strategy by which it wishes to detach itself from the global climate regime partly so that it is not dictated by the international agenda of climate trade and commerce. At the same time, it has also accepted the reality that the best way forward would be to establish close links with the international regime and get whatever it can get hold of. One has to remember that any international policy that is based on the principle of equity at the global level should also be applied to the domestic level. If the per capita principle is applied to emissions, it should also be applied to the distribution and use of energy. At the end of the day, for developing countries, adaptation would be higher up on the ladder of priorities for policy-making since mitigation is expected to have negligible impact on climate change at least in the next two decades.

The politics and science of climate change are intermingled to such an extent that the fates of Governments and leaders have also been shaped by it. In Australia, the Kevin Rudd Government fell because of its inaction on the carbon trading law front after signing the Kyoto Protocol. In the US, ahead of the Presidential elections in 2012, the potential Republican candidates—Rick Perry and Mitt Romney—have come out openly in public against climate actions. Perry commented, "I think there are a substantial number of scientists who have manipulated data so that they will have dollars rolling into their projects. And I think we are seeing almost weekly, or even daily, scientists are coming forward and questioning the original idea that man-made global warming is what is causing the climate to change."[57] If the future of one of the biggest so-called 'climate sinners' is in the hands of such leaders, then the future of negotiations itself is in jeopardy.

Notes

1. Neil Carter, *The Politics of Environment (Second edition)* (New York: Cambridge University Press, 2007), p. 251.
2. Matthew Paterson, *Global Warming and Global Politics* (London: Routledge, 1996), p. 74.
3. Press Trust of India, "EU, India Clash at Copenhagen over 'Climate Justice'," *The Hindustan Times* (December 12, 2009), see http://www.hindustantimes.com/EU-India-clash-at-Copenhagen-over-climate-justice/Article1-485727.aspx, accessed on February 05, 2011.
4. Matthew Paterson and Michael Grubb, "The International Politics of Climate Change," *International Affairs* (Britain: April, 1992), v. 68, n. 2, p. 310, see http://www.jstor.org/stable/2623216, accessed on August 04, 2010.
5. David I. Stern, "The Environmental Kuznets Curve," *Internet Encyclopedia of Ecological Economics* (USA: International Society for Ecological Economics, June, 2003), p. 2, see http://www.ecoeco.org/pdf/stern.pdf, accessed on August 20, 2011.
6. See, John Tierney, "the Richer-Is-Greener Curve," *The New York Times* (New York: April 20, 2009), http://tierneylab.blogs.nytimes.com/2009/04/20/the-richer-is-greener-curve/, accessed on February 12, 2011.
7. n. 5, p. 7.
8. n. 2, p. 76.
9. Ibid., pp. 83-84.
10. Akihiro Sawa, "A Sectoral Approach as an Option for a Post-Kyoto Framework," *The Harvard Project on International Climate Agreements* (Harvard University: Harvard Kennedy School, December, 2008), Discussion paper 08-23, p. ii.
11. William Hare, Claire Stockwell, Christian Flachsland and Sebastian Oberthur, "The Architecture of the Global Climate Regime: a Top-Down Perspective," *Climate Policy* (London: Earthscan, 2010), issue 5, p. 601.
12. Steve Rayner, "How to Eat an Elephant: a Bottom-Up Approach to Climate Policy," *Climate Policy* (London: Earthscan, 2010), issue 5, p. 617.
13. Email correspondence with Dr. Matthew Paterson, University of Ottawa, on August 06, 2011.
14. Eric S. Blake, Christopher W. Landsea and Ethan J. Gibney, "The Deadliest, Costliest and Most Intense United States Tropical Cyclones from 1851 To 2010 (and Other Frequently Requested Hurricane Facts)," NOAA Technical Memorandum NWS NHC-6, National Weather Service and National Hurricane Center (Miami, Florida: August, 2011), see http://www.nhc.noaa.gov/pdf/nws-nhc-6.pdf, accessed on August 29, 2011.

15. See, "U. S. sees Growing Losses from Extreme Weather," *AlterNet* (August 17, 2011), http://www.trust.org/alertnet/news/us-sees-growing-losses-from-extreme-weather/, accessed on August 18, 2011.
16. n. 2, p. 85.
17. Fred Pearce, "Global Warming 'will Hurt Russia'," *New Scientist* (USA: October 3, 2003), see http://www.newscientist.com/article/dn4232-global-warming-will-hurt-russia.html, accessed on September 7, 2010.
18. Interaction with Nitin Sethi, Environment Editor, Times of India (TOI), at TOI Office (New Delhi), on July 4, 2011.
19. Interaction with Dr. Lavanya Rajamani, Professor, Centre for Policy Research (CPR), at CPR (New Delhi), on July 11, 2011.
20. Mukund Govind Rajan, *Global Environmental Politics: India and the North-South Politics of Global Environmental Issues* (UK: Oxford University Press, 1997), p. 97.
21. Ibid., p. 123.
22. Ibid., p. 98.
23. Navroz K. Dubash, "Towards a Progressive Indian and Global Climate Politics," *CPR Climate Initiative* (New Delhi: Centre for Policy Research, September, 2009), p. 9.
24. n. 20, p. 109.
25. n. 1, p. 253
26. United Nations Framework Convention on Climate Change, see for more information, http://unfccc.int/kyoto_protocol/items/2830.php, accessed on March 13, 2011.
27. Aaron M. McCright and Riley E. Dunlap, "Defeating Kyoto: The Conservative Movement's Impact on the US Climate Change Policy," *Social Problems* (University of California: August, 2003), v. 50, n. 3, p. 349, see http://www.jstor.org/stable/3648857, accessed on March 04, 2011.
28. Christoph Böhringer and Carsten Vogt, "Economic and Environmental Impacts of the Kyoto Protocol," *The Canadian Journal of Economics/Revue canadienned'Economique*, (Canada: May, 2003), v. 36, n. 2, p. 476, see http://www.jstor.org/stable/3131853, accessed on August 18, 2011.
29. Gwyn Prins and Steve Rayner, "The Kyoto Protocol," *Bulletin of the Atomic Scientists* (UK: SAGE, March-April, 2008), v. 64, n. 1, p. 46, see http://bos.sagepub.com/content/64/1/45, accessed on August 18, 2011.
30. Jyoti K. Parikh and Kirit Parikh, "Climate Change: India's Perceptions, Positions, Policies and Possibilities," *OECD* (Paris: 2002), p. 22, see http://www.oecd.org/dataoecd/22/16/1934784.pdf, accessed on February 03, 2011.

31. Hitomi Kimura, Ancha Srinivasan and Keisuke Iyadomi, "Clean Development Mechanism," in Ancha Srinivasan (ed.), *Asian Aspirations for Climate Regime Beyond 2012* (Haryana: Institute for Global Environmental Strategies, November, 2006), p. 38.
32. Ibid., p. 46.
33. Interaction with Shyam Saran, Former Prime Minister's Envoy of Climate Change and Former Foreign Secretary, at Research and Information System for Developing Countries (RIS), on June 15, 2011.
34. Ibid.
35. Neha Lalchandani, "From rags to Copenhagen," *The Times of India* (New Delhi: December 20, 2009), see http://articles.timesofindia.indiatimes.com/2009-12-20/delhi/28060367_1_ragpickers-recycling-door-to-door-collection, accessed on September 06, 2011.
36. Joydeep Gupta, "Copenhagen treat' Drafted by NGOs Matches Indian Actions," *India eNews* (October 8, 2009), see http://www.indiaenews.com/asia/20091008/225684.htm, accessed on September 06, 2011.
37. United Nations Framework Convention on Climate Change, FCCC/CP/2009/L.7, "Copenhagen Accord," 18 December, 2009, see http://unfccc.int/resource/docs/2009/cop15/eng/l07.pdf, accessed on March 19, 2011.
38. Richard Black, "UK MET Office Warns Carbon Emissions must Peak by 2020," *BBC News* (London: December 10, 2009), see http://news.bbc.co.uk/2/hi/8405025.stm, accessed on August 24, 2011.
39. n. 37.
40. Interaction with Nitin Sethi, Environment Editor, Times of India (TOI), at TOI Office (New Delhi), on July 04, 2011.
41. Anthony Giddens, *The Geopolitics of Climate Change* (Cambridge: Cambridge Polity Press, 2009), p. 162.
42. Stephen Minas, "Green Climate Fund: A Bad Time to Talk Finance," *Eco-Business* (June 22, 2011), see http://www.eco-business.com/features/green-climate-fund-a-bad-time-to-talk-finance/, accessed on August 18, 2011.
43. David Batty, "Copenhagen Reaction: Delegates Speak," *guardian.co.uk* (December 19, 2009), see http://www.guardian.co.uk/environment/2009/dec/19/copenhagen-reaction-delegates-speak, accessed on August 16 , 2011.
44. Hyung-Kwon Jeon and Seong-Suk Yoon, "From International Linkages to Internal Divisions in China: The Political Response to Climate Change Negotiations," *Asian Survey* (University of California: November-December, 2006), v. 46, n. 6, p. 847, see http://www.jstor.org/stable/4497211, accessed on August 04, 2011.

45. See, John Vidal and David Adam, "China Overtakes US as World's Biggest CO2 Emitter," *guardian.co.uk* (June 19, 2007) see http://www.guardian.co.uk/environment/2007/jun/19/china.usnews, accessed on August 25, 2011.
46. See, Cleo Paskal and Scott Savitt , "Copenhagen Consequences for the U.S., China and India," http://www.upiasia.com/Politics/2010/01/04/copenhagen_consequences_for_the_us_china_and_india/1537/, accessed on August 16, 2011.
47. Interaction with Shyam Saran, Former Prime Minister's Envoy of Climate Change and Former Foreign Secretary, at Research and Information System for Developing Countries (RIS), on June 15, 2011.
48. James Jacob, "Copenhagen and the 'BASIC' Countries: Some Basic Differences between the Indian and Chinese Emissions," *Current Science* (Bengaluru: Indian Academy of Sciences, August, 2010), vol. 99, no. 4, p. 439, see http://www.ias.ac.in/currsci/25aug2010/439.pdf, accessed on August 14, 2011.
49. Interaction with Shyam Saran, Former Prime Minister's Envoy of Climate Change and Former Foreign Secretary, at Research and Information System for Developing Countries (RIS), on June 15, 2011.
50. United Nations framework Convention on Climate Change, FCCC/CP/2010/7/Add.1, "Report of the Conference of the Parties on its Sixteenth Session, Held in Cancun from November 29 to December 10, 2010," March 15, 2011, see http://unfccc.int/resource/docs/2010/cop16/eng/07a01.pdf#page=2, accessed on April 29, 2011.
51. Robert N. Stavins, "What Happened (and Why): An Assessment of the Cancun Agreements," *Media Features* (Cambridge: Belfer Center for Science and International Affairs, December 13, 2010), see http://belfercenter.ksg.harvard.edu/publication/20618/what_happened_and_why.html, accessed on March 27, 2011.
52. See, "Jairam Defends Nuancing India's Position at Cancun," *The Hindu* (Mangalore: December 25, 2010), http://www.hindu.com/2010/12/25/stories/2010122567091800.htm, accessed on February 06, 2011.
53. Nitin Sethi, "India lost a lot, gained little at Cancun climate summit," *The Times of India* (New Delhi: June 8, 2011), see http://articles.timesofindia.indiatimes.com/2011-06-08/global-warming/29633648_1_cancun-climate-kyoto-protocol-rich-countries, accessed on June 09, 2011.
54. See, "Maldives President Pleads for Climate Action," *China View* (September 23, 2009), http://news.xinhuanet.com/english/2009-09/23/content_12098838.htm, accessed on February 12, 2011.
55. Amy Goodman, "Guardian Environment Editor John Vidal on Wikileaks Cables and U.S. Manipulation of Climate Talks," *Democracy Now! The*

War and Peace Report (New York: December 07, 2010), see http://www.democracynow.org/2010/12/7/guardian_environment_editor_john_vidal_on, accessed on August 25, 2011.

56. n. 23, pp. 9-12.
57. Paul Krugman, "Republicans against Science," *The Hindu* (Mangalore: August 30, 2011), p. 13.

4

Geopolitics of Environmental Change: Options and Opportunities

> "The whole aim of practical politics is to keep the populace alarmed — and hence clamorous to be led to safety — by menacing it with an endless series of hobgoblins, all of them imaginary," and, "The urge to save humanity is almost always only a false face for the urge to rule it."
>
> — H. L. Mencken, Columnist

The discourse of climate change has come a long way but it still has a long way to go before it becomes an action-seeking from being one of constant debate. The objective of the negotiations is to have a common stand on the issue rather than to thin out the issue to such a degree so as to facilitate adoption of a common stand by all nation states. The phenomenon of global warming took the centre-stage due to which large-scale environmental degradation and environmental change were pushed to the periphery and stabilisation of temperature rise became the ultimate goal of all climate change negotiations. Some would say that this is the first step towards addressing the larger problems associated with the environment. Some of the idealists, who wanted the world to change its environmental outlook—from exploitation to preservation—had deliberately spawned the 'Doomsday' phobia. As 'Coalition of the willing'—an animated film about an online war against global warming—says, the war on consumerism is synonymous to the war on global warming. In such a consumerist world, industries are hunting for profits and Governments are charging towards growth. As the film propagates, the world is in dire need of a 21st century culture shift, that of global collaborative culture. At Copenhagen, scientists and environmentalists were sidelined while trade unions, nuclear groups and coal groups dominated the proceedings. This is a barefaced sign of diversion form the core issues of climate change. India's role in the negotiations is pivotal, which was proven by the Copenhagen Summit in which the US' agenda to strike a deal in its interest was very successfully blocked by India's strong diplomatic manoeuvring which brought together all emerging economies. India could do much more than this to pressurise the international community to bring all outstanding issues on the table.

India's engagement with the world is on the ascendancy; China's global reach is increasing day by day; and the West is treading the path it has followed for ages more cautiously than ever before. It is true that geography has shaped international relations. However, if environmental change and climate change could reshape geography, this would also imply change in international relations. The negotiations have sharply divided the world and the divisions existing among the international community have sabotaged the negotiations largely as well. In this context, India's position in the global climate change negotiations, its international cooperative measures as well as the possible trajectory of its policies at the negotiations in this regard would be taken up and explored. An attempt would be made to draw parallels between the climate change/environmental policies of India and China and to assess their feasibility in both relative and absolute terms. Critical issues such as CDM reforms, future of ToT, operational and functional viability of the voting procedures at the negotiations, future of the two-track process, positions of other major powers and groupings at the negotiations, and approaches (top-down versus bottom-up) would be scrutinised through the prisms of objectivity and practicality. Contentious issues such as the debate on black carbon, India's strengths and weaknesses and the legitimacy of the negotiations itself would be analysed from various angles. In the end, the effort would be to come out of the vicious web formed by the negotiations and look at the issue from a fresh perspective. India could be the pioneer of change in terms of environmental management.

Balance of Power versus Complex Interdependence: India and China or India versus China

Neo-realists' 'balance of power' theory and neo-liberals' 'complex interdependence' theory try to explain the international system from two diametrically opposite points of view – the former looks at cooperation from a power-centric perspective (power as the means and security as the end) while the latter looks at cooperation from the perspective of institutionalism (international interdependence). India and China have been acting more or less like rivals but the potential for cooperation between the two emerging nations is immense. From India's point of view, bilateral and multilateral cooperation is one of the best ways to not only get integrated into the global climate change regime in a more fruitful manner, but also supplement its domestic actions. India has signed MoUs with several countries on climate change cooperation. During Manmohan Singh's visit to the US in 2009, the United States' Overseas Private Investment Corporation announced plans to invest more than US$ 100 million for clean energy projects in India over the next five years under the Partnership to Advance Clean Energy. During Barack Obama's visit to India in 2010, the two countries launched a Joint Clean Energy Research and Development Centre and a new cooperation programme in India. The US continues to be the highest investor in India's emerging markets and the US Department of Commerce in tandem with the efforts of the US-India Forum have been successful in connecting US businesses and investors with Indian entrepreneurs in the areas of energy efficiency, solar energy and other renewable technology. In July 2010, the joint leadership of the Super-efficient Equipment and Appliance Deployment initiative[1] bolstered the climate cooperation. Similarly, India has also agreed to launch an EU-India Initiative on Clean Development and Climate Change at the EU-India Summit in 2005. The Department of Environment, Food and Rural Affairs (EU) and the Indian Ministry of Environment and Forests have collaborated to conceive projects that would come up with findings on the impacts of climate change on various sectors and climate modelling.[2] India signed a Climate Change Cooperation Accord with China in 2009 by which they not only call upon developed countries to reduce emissions and provide funds and technology to developing countries, but also to cooperate among themselves on energy efficiency, renewable energy and forestry.[3] As the world gears up to mark the 20th anniversary of Rio, India plans to host the Eleventh Conference of Parties of Convention on Biodiversity in 2012. Climate change therefore, could create a platform for countries such as India and China to set aside their political differences and cooperate for the common good.

China is a classic example of a country, which has taken the lead in the field of Renewable Energy Technology (RET) due to increasing signs of climate change on its land in the form of desertification and floods. Its eastern coast is susceptible to sea level rise, northern mainland is prone to desertification and southern mountainous region is at the risk of glacial recession. The concept of 'economic civilisation' introduced in 2008 is a key component of President Hu Jintao's vision of scientific temper and outlook on sustainable development and his commitment to creating a harmonious socialist society. China is known to have the most aggressive deployment of solar and wind energy. It has the world's largest installed renewable capacity; it has the largest number of hydroelectric generators in the world; and it tops the world in solar power generation. China (US$ 54.4 billion) was ahead of Germany (US$ 41.2 billion) and the US (US$ 34 billion) in terms of green technology investments in 2010. It has the largest wind resources in the world and according to a Global World Energy Council study if it continues to advance its renewable energy projects, wind farms would be able to meet 50 per cent of China's power requirement by 2030. Another report released by the Breakthrough Institute and Information Technology and Innovation Foundation has also found that 'China is now poised to surpass the West in virtually all areas of clean energy, including transportation technologies, high speed rails, advanced batteries, solar water heaters and even next generation nuclear and 'clean coal' technologies'.[4] India could learn a few valuable lessons from China in this respect. For example, China introduced differential taxation to derive profits from non-CO_2 CDM projects, which in turn would be used to create a sustainable development fund and to promote renewable energy prospects. Since the enactment of the Renewable Energy Law in 2006, it has also subsidised the investment, transmission and distribution costs of renewable energy-based power projects. A number of other environmental fallouts of this renewable energy investment drive have counteracted these figures. For example, a solar panel factory had to be shut down in 2011 after protestors demonstrated against the plant for polluting a nearby river in the eastern Chinese province of Zhejiang.[5] Similarly, the Three Gorges Dam has been held responsible for 'risk of geological disaster' as it is linked to 'soil erosion, quakes, drought and social upheaval' as the State Cabinet itself has admitted.[6] The drying of major rivers of any country defeats the purpose of setting up of hydroelectric projects on them as they would not only have disastrous environmental consequences but also lead to shortage in the production of electricity.

One of the characteristics of China's 'peaceful rise' has been its endeavour to control environment. One of their hydraulics engineers, Professor Liu Zihui, while speaking about the largest hydraulics project in the history of humanity, a US$ 63 billion canal that would bring water from the south of China to the increasingly desertifying North China, commented, "I don't feel we are conquering nature. We think nature itself isn't fair. God isn't fair. What is that? He's given Southern China so much water but given Northern China so little. It's a good land—nice flat land—up there. But it's got so little water. So we say, as God isn't fair, we are trying to balance out God's unfairness."[7] Here is a classic example of a country that has been upholding the bottom-up approach at the international level but at the domestic level, it has centralised the entire environmental governance system like any other aspect of the Chinese system, thereby not losing vision of the long-term interests of the country. India, on the other hand has been following a more or less bottom-up approach both at the international and domestic levels. India should not follow China's example by trying to control environment; the need of the hour is to manage environment at the micro-level as well. Even rainwater harvesting schemes run by individual houses could be solutions rather than large-scale projects that could only solve the problem temporarily. The North-South Canal is one of the treasured projects of China that it claims could solve the water crisis and address the water-related problems of climate change such as flooding in the South due to glacial melting. India is also thinking on similar lines to interlink all rivers of the country to maintain water security. However, such large-scale projects could be severely affected by environmental change and climate change, which could only exacerbate the water security situation. India is the storehouse of a number of traditional technologies and methodologies, which could be promoted elsewhere in the world.

Geopolitically, ties between China and India have been tense. Tensions along the border have now extended to the Indian Ocean over China's growing relations with countries in the region and even to the South China Sea over India's ambitions to explore oil and gas in the region. In such a situation, technology cooperation between the two countries seems highly unlikely especially since the two are also competing among themselves in a 'green game' in Asia even though they have had similar postures at the negotiations. Sino-Indian cooperation in RET can be boosted only by private players as the Government of India would be highly reluctant to engage with a country, which has menacing regional as well as global aspirations; added to that is the historical baggage of war and trust deficit. According to Shyam Saran ,the West is still ahead of China in terms of cutting-edge technology. In addition, familiarity between the Indian entrepreneurs and the Western ones is more. Bilateral dealings are different from multilateral dealings. At the end of the day India has to build its own capacity.[8] Others would suggest that India should look forward to cooperating with every country including China despite political hostility as South-South cooperation on environment and development is the key to future discourse on climate change and sustainable development.

Niu Qingbao, Consul General of China in Mumbai, said that the level of cooperation between India and China is very low currently even though the potential exists and that there are no obstructions whatsoever from China's side.[9] However, India's cooperation with the industrialised world itself has not been satisfactory. There have been instances in which the West has shown its willingness to pursue joint research and development in renewable energy as their economies are in bad shape. But India has not shown the political willingness to go down this path in the manner it is expected to at this stage; this would make it completely depend on the West in future. It has to learn from its lessons in defence in which India has now initiated joint R&D that would result in speedy outcomes.[10] Interestingly, environmental diplomacy has become an integral part of the struggle for supremacy, so much so that a country's expertise in environment and climate in various sectors could be a potential soft power tool just as public diplomacy is. China's entry into the scramble for resources in Africa after the exploitation of the continent for more than two centuries by the Western powers has raised serious concerns in some of the countries even though moves are being made by China to address environmental challenges through the Forum for China-Africa Cooperation. The Western powers are now dictating terms to developing and underdeveloped countries while providing them aid to develop low-carbon infrastructure and technologies.

Recently, even a country such as New Zealand offered Tonga aid for a solar power plant on the precondition that the plant would be built only by a New Zealand government-owned company, which goes against the principles of no tied aid. If there is no competitive bidding, the costs of energy are expected to rise, which the economy of Tonga might not be able to withstand in the long-term.[11] India, on the other hand, has a better international image in terms of environmental footprint abroad. Its knowledge pool in this field is considered one of the largest and thus, it should pursue a more vigorous environmental diplomacy to enhance its global reach. Environmental diplomacy, if utilised in an equitable fashion, could improve relations between countries, and lead to international peace and stability.

The Case Study of Black Carbon

The case study of black carbon would set the agenda for a healthy discussion on how India could demarcate its domestic policy from its international policy. It would also facilitate a better understanding of the complexity of the issue. On the question of black carbon or soot, India has to tread the path shown by the West carefully. It is emitted into the atmosphere because of incomplete combustion of fossil fuels. Black carbon has been pinpointed as the single biggest factor in the formation of the Asian Brown Cloud. This is nothing but a layer of air pollution that covers parts of South Asia mainly between the months of January and March. Its chief generators are diesel engine exhaust and cooking fires. As the black carbon to sulphate ratio increases, the absorption of solar radiation has been found to increase. At the same time, it has also been found that while the atmosphere gets warmer, the surface of the earth experiences a shading effect by which the earth's surface cools. A series of studies have revealed that black carbon could be responsible for nearly 18 per cent of global warming, which makes it the second biggest factor after CO_2, and India is known to be contributing more than 6 per cent of the world's black carbon emissions.[12] While the life span of CO_2 could be several centuries, the life span of black carbon is usually only couple of weeks. Black carbon has not been one of the focuses of the IPCC and in fact, Dr. R. K. Pachauri himself had dismissed the linkage of black carbon with global warming and climate change back in 2009[13] when there was mounting pressure from developed countries to include this component in the negotiations agenda apart from the other GHGs at Copenhagen. But pressure from the Western scientific and political communities is at its highest at this point in time; this is one of the reasons for the launching of the Black Carbon Research Initiative under the aegis of the Indian Network of Climate Change Assessment by India's erstwhile Minister of Environment and Forests, Jairam Ramesh in 2011. This can be treated as another sell-off on the part of the Minister while others would regard it as a proactive step to enhance India's study of climate change and its effects.

It is the impact of black carbon on glaciers that entails maximum research as its warming effect is found to be greatest in snowy places such as the Arctic and the Himalayas due to the albedo effect. If black carbon settles on the ice and snow, it would absorb most of the solar radiation with very less light being reflected back into the atmosphere, leading to the melting of ice and glacial loss. If one looks closely at the phenomenon of black carbon in India's context, one has to remember that the number of people living on biomass has remained constant since 1950s and 60s due to the spurt in urbanisation. The other cause of black carbon—diesel exhaust engine—despite proliferation of diesel vehicles has also been neutralised by rapid improvements in diesel technology over the past thirty years. The new technologies not only utilise the fuel effectively but also give better mileage. As far as the claims of the scientists are concerned, these are mainly based on models, which make use of statistical tools. So, the credibility of scientific proof can be questioned on various counts. The fact of the matter is that the alternative to biomass burning for cooking is Liquefied Petroleum Gas (LPG), which is known to emit more CO_2. Therefore, in the long run it is a more unfriendly solution for cooking.[14] The question of whether or not black carbon causes global effects again needs to be re-looked at and freshly investigated. Since all the scientific studies on black carbon have been conducted by the Western scientific community, the veracity of their claims needs to be checked by the scientists in India and China, which account for more than one-third of the black carbon emissions. Moreover, particles have to stay in the atmosphere for a longer period of time to trap heat. The argument that the Western scientific communities' claim is nothing but an exaggeration and an attempt to put pressure on developing countries such as China, India, South Africa and Brazil to take action is compelling. According to them, since black carbon remains in the atmosphere for a limited period of time, it is the easiest to attack or curb. Addressing the black carbon issue by removing it could bear faster results. At the same time, India and China will have to study the localised impact of black carbon, for example on the Himalayan glaciers or on human health, which are of greater significance.

The Road to Durban (COP-17): Recommendations and Growing Uncertainties

Technology is the Buzzword

Though it was at Bali that the need for technology transfer was given the final push, it was the Cancun Agreement that made formal provisions for a Technology Executive Committee, a Climate Technology Centre and a Climate Technology Network. Movement of technologies from the developed to developing countries has been obstructed by the former for a long time for various reasons including fear of losing hegemony in the technological market on account of reproduction of technologies by the recipient countries. Instead, they have shown more willingness to sell or provide in aid outdated technologies to developing countries. The proposed network should not restrict itself to technology transfer in the context of North-South dialogue; it should be extended to South-South cooperation since countries such as China and Brazil have already taken huge leaps in this direction. Along with strengthening and redefining the roles of the existing technological institutions, new centres have to be set up to meet more pressing demands especially in the case of the Least Developed Countries (LDCs) and the Small Island Developing States (SIDS). The need for a robust association between the technology mechanism and the financial mechanism is a prerequisite for the former to take shape and function effectively. Not only should the technology mechanism be funded by the financial mechanism, but also it should be able to provide inputs to the latter in terms of cost-benefit analysis as well as futuristic assessment of the technologies. Some of the ways in which technology transfer can be made more efficient and even-handed are:

- Not only should the decision-making bodies comprise of countries from both the developed and developing/underdeveloped blocs but also it should operate on equal terms and not based solely on financial contribution.
- Technology cooperation should be given a boost apart from technology transfer so that the relatively financially well-off developing countries need not depend on developed countries for cutting edge technologies besides the fact that the equations of competitive advantage would serve both blocs' interests.
- The funds for technology research and development and transfer, have to be secured and in this respect, the contribution of developed countries has to be substantial and sustained. Some would suggest this could be implemented by making developing countries accountable for their actions or even imposing a global carbon tax with no deviation from the 'common but differentiated responsibility' since uniform global carbon tax is unarguably politically and economically unethical.
- The needs of different countries could be different and therefore an appropriate analysis of the needs has to be done before the deployment of technology. In such a scenario, both adaptation and mitigation have to be given equal focus. For example, the majority of Asian and Pacific countries require adaptation technologies in the areas of agriculture, water resources, coastal zone protection and forest management.[15]

- The issue of Intellectual Property Rights has to be resolved in order to allow the recipient countries to replicate the technologies as the issue is a global challenge and every country is a stakeholder.
- Autonomous scientific appraisal, public-private partnerships in technology development, involvement of the NGOs and the civil society in the implementation of projects as well as establishment of a feedback system are essential.
- It is very important for countries to refrain from indulging in malpractices and corruption such as funding of inappropriate technologies which may cater to business or political interests and not environmental or human interests.

Let's Make Clean Development Mechanism Cleaner

Another very important tool for the development and deployment of technology is the CDM. The system of the CDM itself needs to be revamped in order to include various methodologies that are not currently recognised under the CDM instead of a project-based CDM, which limits its scope and efficiency in the long run. It is true that the Kyoto Protocol's future itself is in a quandary as of now and therefore, the CDM could be in jeopardy. However, the chances of it surviving are decent especially since it has become an integral part of the global carbon market along with the CO_2 emissions rights and no country would like to sacrifice the financial proceeds that they gain from more than a US$ 100 billion market. However, its name and scope could undergo transformation. Lack of clarity about the existence of the market in the post-2012 scenario besides recession (due to which emissions of industrialised countries reduced naturally) have dampened its growth to some extent in 2010 and 2011 according to the World Bank. "The global carbon market is at a crossroads. If we take the wrong turn we risk losing billions of lower cost private investment and new technology solutions in developing countries," said Andrew Steer, the World Bank Special Envoy for Climate Change.[16] The CDM needs to be given an impetus by replacing the project-based system with a policy-based one, which allows 'public policies aimed at reducing GHG emissions to be eligible for the CDM without pre-established limitations in terms of geographical coverage'. A better option would be the sectoral CDM in which 'GHG emissions reduction activities along the lines of a sector or sub-sector are made eligible for the CDM project regardless of the type of enabling instruments',[17] since it would be easier to implement it rather than a policy-based one which could encompass several sectors making the system more complex. For example, in the transportation sector, this would ensure that activities such as increasing fuel efficiency and the choice of mode of transport or even avoidance of transport would be counted and given credits. One must also remember that the SIDS and the LDCs probably could not be kept at par with China and India; they would have to be integrated with the CDM in a much bigger way. Their dependence on Annex I countries in the long-term in the form of fast-track finances could affect their economies and self-sustainability. The Certified Emission Reductions (CERs) allocated to these countries should be of a higher value as compared to actual CERs so that equity is maintained.

Beyond Copenhagen: Light or Darkness?

The post-Copenhagen scenario has left several questions unanswered despite a so-called 'successful' Cancun Agreement. Two conferences were held in 2011 before the COP-17 at Durban. One was held at Bonn from June 06-17, 2011 and the other was held from October 01-07, 2011. The signals that these conferences have sent are not very positive. The most prominent among them is the continuing bickering over finance and aid. It is not clear, at this juncture, if consistency and sustainability of financial assistance would continue after 2012 when the first commitment period of the Kyoto Protocol ends even though developed countries have made pledges at both Copenhagen and Cancun. Developing countries, the LDCs and the SIDS have agreed to be more transparent about the use of the international aid for adaptation and mitigation but the process of collection and disbursement of aid has been dead slow. At the Panama Conference, while the US refused to discuss about the long-term climate finance as according to it, the issue was already settled at Cancun, developing countries insisted that with the end of fast-start finance in 2012, the uncertainty would continue with respect to long-term finance.[18]

On one hand, countries such as the US and Canada have failed on the mitigation front completely in the last one decade. On the other hand, the BASIC countries have expressed their readiness not only to adopt legally binding emissions reductions targets at a legally appropriate forum but also to be transparent with respect to their domestic actions. The EU has made considerable efforts to address the issue and with its historical tradition of propagating universalisation of the European ideals, it would be delighted to take the lead on this issue too. The onus now lies on the BASIC countries to apply pressure on industrialised countries to push for stronger mitigation efforts on their part. Consolidation of the two tracks, that is Ad-Hoc Working Group-Kyoto Protocol (AWG-KP) and Ad-hoc Working Group-Long Term Cooperative Action (AWG-LCA), is being seen as one of the options. Another option that is being discussed widely is whether to continue with the principle of 'consensus' or to switch to the 'majority' principle. Germany introduced the issue in the United Nations Security Council in July 2011 to push for a deal at Durban just as the UK did in 2007 to push for a deal at Copenhagen. UK failed; Germany's chances of victory remain contingent on the actions of the Umbrella Group (UG) and the BASIC. Australia and Norway, have meanwhile called for delaying a legally binding global climate treaty until 2015 by submitting a joint proposal to the UN.[19] These developments highlight a bleak future for the global climate change negotiations—(a) countries trying to push their interests by securitising the debate only to gain leverage in the global carbon market, (b) countries trying to stick to the current discourse on climate change and gaining little both domestically and internationally, (c) countries selling themselves off to the rich parties for short-term gains, and (d) countries refusing to initiate negotiations on the post-2012 regime and using delaying tactics to reduce the burdens on themselves. In this situation, the majority-based approach might make more sense. Two main scenarios could be built in this context:

- **If the consensus-based approach is continued**

As already discussed in the earlier chapter, dissent by any particular country has to be taken into account while formulating the draft proposal. In this respect, every country should be treated equally and not differentially when it comes to adoption of the agreement with due consideration to every country's arguments for or against it. In practice, this approach has been hardly observed. A group of countries has handled the drafting of the final agreements in most of the cases with the other counties endorsing it for the 'common good'. Therefore, the UNFCCC should now rethink over the practicality of this approach.

- **If the majority-based approach is adopted**

This approach could be handy in terms of arriving at stronger agreements on reducing emissions and flexibility mechanisms. The UG countries and the Organisation of the Petroleum Exporting Countries (OPECs) have literally rejected the idea of a Kyoto-type agreement after 2012 that according to them harms their interests. It is high time that these parties are overridden in order to block their delaying tactics and incorporate sound policies. At the same time, one could also argue that this could encourage the formation of the 'coalition of the willing', which would kill the negotiations. Climate change is a global phenomenon and it has to be dealt globally. Experts have suggested that countries could be prevented from blocking the progress of the talks by having 'three quarter majority voting, double majority voting on certain issues'. There is also a move to adopt the 'friends of the chair'[20] approach which would mean that a group of participants would be selected to come up with an agreement which then would be imposed on the rest of the parties. This could prove to be a disaster in the light of the international community's experience with the 'veto' of the UN Security Council (UNSC). Creation of an exclusive club would certainly not solve the world's problems; it could only aggravate them.

As far as the future of the two-track process is concerned, there could again be two scenarios:

- **If the two-track process is continued**

The two-track process is contingent on future of the Kyoto Protocol and role of the US. If the US chooses not to have any internationally binding targets, the chances of other developed countries following suit are high. The second commitment period therefore might not come into effect. The centre of the debate is yet again the US since it has not joined hands with the other developed countries to fulfil its binding targets under the Kyoto Protocol. There is a feeling that if the two-track process is abandoned and emerging economies such as China, India and Brazil are brought under the ambit of a binding international agreement, the US might agree to ratify it. However, as already highlighted in the earlier chapter, the Cancun Agreement text itself blurs the differences between the developed and developing countries' commitments.

- **If a single negotiating track is adopted**

Consolidation of the two tracks has been a long-standing demand of the UG countries particularly the US according to which the issue requires both developed and developing countries to come together and negotiate at the same level. This could be the biggest setback for developing countries, as a single negotiating track would undermine the 'common but differentiated responsibility'. If the single track eliminates the existing differences between the developed and developing countries, the former despite having far more financial and technological resources will get away with low commitments in relation to the latter, which would be forced to take up burdensome commitments. However, countries such as the US that have been non-committal since the beginning could exploit the single track in order to tilt the negotiations in their favour.

The US is currently part of the AWG-LCA while all the other developed countries are part of the AWG-KP. It has been pushing to formalise the 'bottom-up approach' so that it does not have to take up binding international targets and at the same time, domestic targets of the advanced developing countries would also become legally binding as well as the domestic actions of the most vulnerable countries would be measured, reported and verified (MRV) internationally. Therefore, the second commitment period would mark a sea change in its approach towards mitigation of climate change. First, the line between the developed and developing countries would gradually become more and more irrelevant. Second, the international approach would be replaced by a domestic approach that would dilute the significance of the UNFCCC. Third, the financial mechanisms might become redundant. The issue of compliance has already been a non-starter. The enforcement branch of the mechanism states that if the emissions of a Party exceed its assigned amount, 'it must declare that that Party is in non-compliance and require the Party to make up the difference between its emissions and its assigned amount during the second commitment period, plus an additional deduction of 30 per cent. In addition, it shall require the Party to submit a compliance action plan and suspend the eligibility of the Party to make transfers under emissions trading until the Party is reinstated'.[21] If the second commitment period of the Kyoto Protocol does not come into effect, these words would never be translated into action. Even if it does, the present non-complying parties such as Canada could pull out of the Convention and not pay any price for its non-compliance during the first commitment period.

According to the popular belief, the UG countries are ready to sign an international agreement if and only if developing countries are brought on the table to discuss about future binding emissions cuts. Countries such as Australia and Japan are also keeping an eye on the US' moves, and its future course of action with respect to emissions cuts. The European contingent has always portrayed a very 'progressive' image at the negotiations with fresh targets and proposals to reduce emissions. The reality seems to be the exact opposite. As Matthew Paterson states, "It seems to me that industrialised countries could live with China and India having something like 'no lose' targets or intensity-based targets. Indeed the more optimistic parts of the run-up and fallout from Copenhagen and Cancun involve this effectively— China's target for example was accepted as legitimate by the industrialised countries. The broader question seems to me really about when specific countries internalise a 'win-win' argument about emissions reductions—the big problem in North America for example is not so much the developing country problem, but that they still see reducing emissions as a big burden (unlike say the UK or Germany, or for example South Korea)."[22] This shows that even if developing countries adopt binding emissions reductions, the Umbrella Group countries would still find ways to bypass any binding agreement.

The European contingent is now perpetually divided on the issue with the EU going in one direction and countries such as Norway and the UK (despite being a member of the EU) seems to be heading in a diametrically opposite direction. Latest reports suggest that the UK Government is in two minds about its ambitious emissions cuts target of 50 per cent below 1990 levels by 2025, as the Conservative Party has significant number of dissenters against the Government's 'green rhetoric', who hold the country's environmental laws responsible for 'piling costs on to energy bills'. From its 'vote blue, go green' slogan—a major electoral issue that had dominated the last election campaign—the Tories have started to downplay environmental issues.[23] The focus has been more on nuclear energy than renewable energy despite the fact that the IPCC has established through its findings that renewable energy is capable of meeting 80 per cent of the world's energy demands by 2050.[24] This is the reason why countries such as India and Japan have been calling for inclusion of nuclear energy in the CDM, even at the Bonn and Panama Conferences.[25] This could give countries such as the US, Russia and France a big boost since they are three of the biggest producers of nuclear energy and this would mean that they would be allowed to emit more than what they are currently emitting. Therefore, the reality is that with nearly all countries in the UG are willing to sign an international agreement if they are given 'space' to emit more and more.

The top-down versus bottom-up approach debate continues and time has come when the international community including the research one should now concentrate on the process of consolidation of these two approaches both at the domestic as well as the international levels. True and complete environmental governance could be achieved only if both the approaches operate in cohesion simultaneously. Even if small communities and groups could create self-governing systems of environmental management, it should be mainstreamed so that other communities emulate these patterns across boundaries, which could then be incorporated in global climate change policy. To say that compliance mechanisms could work only in the case of a top-down model is a farce. For example, the Kyoto Protocol is considered a top-down model with 'binding' targets but due to loose compliance mechanisms, countries such as Canada which would certainly not be able to fulfil its commitments at the end of the first commitment period would still get away with no penalties but mere image destruction. The diversification at the UNFCCC level itself has been recommended by several legal experts on climate change, that is, the portfolio approach in which the Kyoto-based approach for the existing parties, a hybrid of top-down and bottom-up approaches for the non-Annex I parties and pure bottom-up approach based on only national actions for the LDCs would be applied.[26] This is the 'ideal' approach towards tackling climate change but in reality, a significant number of parties are not even willing to go through the legal process to replace any of the existing mechanisms with stronger ones. Until now, a majority of legal principles have been violated by countries such as the US that ratified the UNFCCC but has failed to ratify the Kyoto Protocol as according to it, the world has changed since 1992. Therefore, the challenge is to implement the existing legal instruments first and then look for future probable ones that should not be allowed to fail. Matthew Paterson points out, "The more important question is how you design an agreement so that countries comply with it by choice."[27]

The Strategy-building Exercise in Tandem with the Rethinking Exercise

By now, it is very clear that the climate change negotiations have virtually reached a deadlock. The majority of decisions made at the Cancun Summit have not been implemented and no country has the political inclination to lead the negotiations in its current form or even refined form as the state of affairs might change in future. Pessimism, which has been a hallmark of the negotiations since the coming into effect of the Kyoto Protocol, is expected to continue at least for the next two-three years. The obsession with 'making money' among certain countries has made the negotiations seem like trade negotiations, which fundamentally follow the model of the World Trade Organization's negotiations. This has led to the entire focus of discussion being shifted to the 'cost' factor that overshadows several benefits of being environment-friendly. This is the exact reason why the negotiations have lost their credibility and the only way in which some life could be breathed into the negotiations is to take a few steps backward and rethink over the international community's approach towards climate change. India has to deal with the climate change negotiations and the major countries in the world from a position that lays emphasis on the 'right' issues and does not allow the attention to be shifted from true 'national interests' as well as 'international peace and stability'. The questions that need to be re-looked into are:

The Old Question	The New Question
Is climate change 'natural' or 'human-induced'?	How does the world deal with the challenges posed by changes in climate and environment—natural or human-induced?
Which countries are responsible for climate change?	Are climate change and environmental change a 'developing country syndrome'?
Which countries should pay the price for adaptation and mitigation around the world?	How ready are the countries that are 'responsible' for climate change to improve resilience against both climate change and environmental change in their own countries?
On the front of mitigation, how does the international community reduce carbon emissions that are largely held responsible for global warming and climate change?	On the front of mitigation, would reduction in carbon emissions alone address the challenges posed by climate change and environmental change?
On the front of adaptation, how does the international community protect the most vulnerable countries?	On the front of adaptation, how does the entire international community beef up its infrastructures—physical, economic, legal, political and social?
Is climate change a security threat?	Is environmental change a security threat?

To elaborate on these observations, the existing literature on possible strategies could be taken as a reference point and the same strategies could be tweaked a bit to strengthen India's strategy based on reality and practicality. The existing ones lay stress on the per capita emissions principle, moral and ethical principles involved in the negotiations, the 'common but differentiated responsibility', technology transfer, financial aid and so on. It is high time that countries such as India adopt these pillars of negotiations only to facilitate the transition of talks from trade-related issues to environment-related issues rather than portraying themselves as stumbling blocks in reaching a global climate change agreement. Even when the debate is securitised, countries such India, China and Russia have opposed such moves purely based on its reservations about the Western agenda of pushing a 'trade agreement' in its favour. This message has to be sent across to the international community loud and clear, or else the negative image of India as a 'stonewaller' would never get eliminated.

This however requires countries such as India to come forward with fixed standpoints on the issue and bring to the attention of the international community about the lopsided nature of the regime. It has to convince the other developing countries and the amicable developed countries that the way forward would be to concentrate on environmental change rather than mere climate change so that the threat of human-induced environmental change to developed countries could be underscored. Due to the over-emphasis on carbon emissions because of business interests attached to the cap-and-trade mechanisms, the real issues that face countries such as India are being ignored. The need of the hour is to expand the international community's understanding beyond carbon emissions and look at other GHGs as well besides other concerns such as groundwater depletion, land degradation, overfishing and deforestation among others.

Developed countries must get rid of their 'developed country complacency syndrome' and address the issue holistically. The West has been adding fuel to the issue of climate change only to put pressure on emerging economies to halt their carbon-based growth. However, they themselves have been hardly paying attention to the threats posed by environmental change to their own countries. For example, the havoc caused by hurricane Katrina was exacerbated by a number of human-induced factors such as faulty levee design and implementation, lack of town planning and large-scale subsidence caused by draining of wetlands, extraction of groundwater and inappropriately designed waterways.[28] In yet another example, Lake Mead in the middle of the Nevada Desert in the US, one of the biggest man-made reservoirs in the world whose water is held back by the Eighth Wonder of the World—the Hoover Dam—has been witnessing a decline in the water levels since the beginning of the twenty first century. The causes have been several and are not restricted to just climate change; they include multi-year drought and continued growth especially in terms of population in the Southwest. Cities including Las Vegas, Denver, Los Angeles and Phoenix and other areas in the states of Nevada and Arizona would face serious water shortage apart from possible power shortage if the water levels dip further that would disable the Hoover Dam's hydroelectric turbines.[29] In fact, even in tackling climate change, the other GHGs such as methane, nitrous oxide and the most important factor, water vapour have been discounted. If one can break the debate down to issues such as methane emissions, an area that a country such as India could easily capitalise on, countries such as the US would be forced to come to the table. For example, one could argue that India should get credit for the overwhelming number of vegetarians in the country in comparison to the predominantly meat-eating West. In addition, the countries in the West should be penalised for using toilet paper which entails deforestation—a practice which is not common in countries such as India. The cattle industry is subsidised in the US. If one reflects on the impacts of cattle industry, one could easily say that the West needs to clean its backyard before pointing fingers at developing countries. The water utilised to produce one pound of meat amounts to 2,500 gallons. More than half of the US agricultural land is used to produce meat, and this leads to top soil erosion and depletion of forested areas.[30] Moreover, methane and nitrous oxide emissions which are considered twenty times and three hundred times (respectively) more potent than carbon emissions are mainly produced by the livestock and meat industries.[31] The Indian livestock and meat industry is a dormant one in comparison to the ones that thrive in the West. Therefore, India should put pressure on the West to consider these important factors while negotiating a climate change deal, which currently is focussing on curbing only carbon emissions. This could embolden the blame game that has already consumed enough time and resources on the climate change issue. Yet, the need for such an avant-garde debate is highly relevant in international relations and this is the only way in which the industrialised and emerging countries could be brought to the board, and forced to negotiate on equal and practical terms.

It is very clear that the global climate regime is highly biased due to its inherent economic and trade undertones. The energy security debate runs parallel to the climate change debate. The question that India has to ask is that how much should the country take the climate change-energy security nexus seriously, and how much should it concentrate on science. The negotiations have well bypassed the science of climate change debate and jumped directly into the energy security debate. Therefore, it is imperative for countries such as India to keep an eye on the science of climate change while being active at the negotiations. In order to differentiate between the two, India's nuclear energy policy can be taken as a classic example. India could adopt nuclear energy as the bedrock of its energy security policy and set up as many nuclear power plants as possible across the country. At the same time, the Environmental Impact Assessment has to be carried out in such a manner that not only the impact of the project on environment but also the impact of environmental change on the project should be assessed minutely. Any sudden changes in climate or environment could affect the nuclear power plant drastically that could threaten not only energy security but also human life and other physical infrastructure. Therefore, the thrust should be laid on protecting the existing infrastructure from the environmental changes as much as on the need for meeting the energy requirements that deviates attention from the core of the climate change debate. Developing countries including India should understand that it is not enough to reduce use of fossil fuels to prevent climate change and extreme weather events. They might occur irrespective of whether or not climate changes over a period of time. Therefore, human security should be on top of the agenda.

India should, therefore, participate in the negotiations and attain whatever they can get hold of in terms of financial and technological assistance. It is very clear that there can be no global climate deal without India's participation. Hence, India should use its influence at the international level to acquire a deal in its favour. It has to deal with international politics and economics of environmental and climate change differently at both international and global levels. At the international level, it should continue to underscore the per capita emissions principle to ensure that no unfair regime is imposed on it. On the negotiating table, it has to learn a few lessons from China, which agrees to comply with all international norms, but is seen to be in contravention of those very norms as evidenced by its nuclear cooperation with Pakistan despite being a signatory to the Nuclear Non-proliferation Treaty (NPT) and Pakistan's dubious history of proliferation. India should not approach the negotiations with a black band and confront isolation, and pressure to toe the line of the West. It should try to engage with the rest of the international community including the West in such a manner that it is not labelled an 'encumbrance' on the one hand and 'moral' and stupid' on the other. At the domestic level, it should not commit the mistake that China is currently indulging in by going on an 'infrastructure-building' rampage that is supposedly 'green' but could prove to be disastrous in the long run. India has to beef up its environmental security and the physical and legal infrastructure associated with it besides reducing its carbon emissions.

It would be too ambitious to expect industrialised countries at this stage to cooperate with developing countries and share the costs and benefits as they are facing financial crunch. At the same time, one must not forget that India has a number of small-scale traditional technologies that it could promote elsewhere in the world especially among developing countries. Technology cooperation among developing countries and even between the developed and the developing countries requires political will. It is very clear that countries such as Germany and the US are on their way to make not one but several technological breakthroughs, particularly in RET. It is also clear that pressure would mount on developing countries to buy these technologies in future when an international agreement is reached upon. Therefore, it is crucial for India to negotiate at the international level and frame environmental policy at the domestic level in such a manner that its future is in its hands itself and not in the hands of developed countries. A technological breakthrough in renewable energy in India that it could sell at a lower price than what the West has been offering to the rest of the world could prove to be a game-changer for India.

The need of the hour is to exhibit India's potential to exercise its soft power and develop new technologies to be self-sufficient. For this, coordination between the Ministry of Environment and Forests (MoEF), the Ministry of Earth Sciences, the Ministry of New and Renewable Energy and the Ministry of External Affairs (MEA) has to be wholesome. History has shown that differences between the MoEF and the MEA have derailed India's negotiating positions on various occasions including the Copenhagen Summit in which reports insinuated that Shyam Saran (Prime Minister's Envoy of Climate Change) and Jairam Ramesh (the Environment Minister) were at loggerheads. Shortly after the Summit, Shyam Saran resigned from his post, and the Prime Minister invested the Environment Minister with complete authority as far as climate change negotiations are concerned. Even to this date, differences are conspicuous. Bureaucrats admit that though systems are in place, there are differences that need to be sorted out.[32]

India's emissions are only expected to increase in the coming years. The chances of them peaking in the next twenty years are bleak as evidenced by several estimations by the International Energy Agency. The principle of equity at the international level should not water down to business-as-usual at the domestic level, which may not prove to be beneficial for the country in the long run. At the same time, concessions made by India at the international level have to be neutralised by the fulfilment of the developing countries' demands of financial assistance as well as a more holistic approach to the issue by developed countries. This principle has to be broadened to include the climate change or environmental change-related security threats for developed countries. The word 'developed' does not indicate immunity from the vagaries of climate change. It does not imply absolute readiness to improve resilience either. Therefore, the attempts by developed countries to portray the climate change problem as a problem of developing countries has to be countered. This could be done in two ways. First, developed countries should be pressurised to broaden the debate to environmental change. Second, the notion that reduction in the consumption of fossil fuels, which in turn would lead to reduction in the GHG emissions, reduction in the effects of climate change, reduction in the extreme weather events, has to be discarded so that there is 'consistency' in terms of identification of the causes of and response to climate change. One of them is adaptation.

In fact, one could say that climate change has shifted focus from environmental change and this is the reason why the climate change negotiations have been deliberately turned into trade negotiations. Even if the emissions reduce to zero now, the impacts of climate change would still be felt for centuries because of accumulation impact. Therefore, both developed and developing countries should address the issue of adaptation, which could serve the whole world and humanity's interests in the coming years. Consistency can be infused into the global climate change system only if developed countries are convinced that they are equally endangered. The West, especially the US has been violating the principles of equity and consistency within its own domestic constituency. There are several examples of the flip-flop attitude of the West. The emissions trading mechanism has allowed several countries in Europe to trade their emissions with the economies in transition whose emissions have been far low than their targets and get away with a business-as-usual scenario. Second, by examining the impacts of climate change alone, the West has ensured that only trade-related aspects of climate change including emissions are addressed. At the same time, countries such as the US have been making a mockery of their own policy on climate change by neither reducing emissions nor protecting their physical infrastructure. Therefore, contextualisation of equity and consistency—two concepts that have been exploited by developing countries to press their demands on industrialised countries—has to be fine-tuned to add more aspects related to the climate change debate.

In Brajesh Mishra's words, "India should adopt a two-pronged strategy – first, to assert that India would not give up its sovereignty, and second, to educate people especially in the US that climate change would affect them and not just the developing countries."[33] In short, India's approach to climate change negotiations in future should consist of the following three tenets:

- It should take the responsibility of dealing with climate change and environmental change and thus take the responsibility for its own security rather than waiting for foreign assistance.
- It should continue to stress on the per capita emissions principle, which seems to be the only long-term legitimate principle.
- It should negotiate for whatever it can get in terms of technological and financial assistance without giving up its core interests.

Do We 'Breathe in' or 'Breathe out'?

It is certainly in India's interest to negotiate sagaciously and resiliently. If India concentrates all its policies to the domestic circle with no or little regard to the international scene, its isolation is imminent. But more importantly, India's engagement with the rest of the world that is of great consequence for its 'great power' status or 'great power in the making' status would be affected severely. It is not in India's interests either to concede its ground to its competitors or to develop antagonistic attitude towards them in international relations. It is imperative for India to secure its environment and population from possible negative developments—natural or human-induced. Therefore, just as it is important both to breathe in and breathe out to survive, it is important to harmonise a country's internal and international policies in such a manner that they are separate operationally but similar functionally—in supreme 'national interest'.

The financial mechanisms have failed to deliver any fruitful results so far. The future of the Kyoto Protocol still is uncertain. The scope of the CDM stands where it was a year or even two years back. The US has not budged an inch on the climate change issue though the Democratic Government still sends its representatives to the summits and conferences only to stall the negotiations and add to its quota of GHG emissions caused by its representatives' air travel. The question of equitable energy and technology distribution as well as sustainable development at the global level has been left open for years and years of debate and no action even though energy security has been tied to climate change so intricately. The negotiations have reached a stage in which any target fixed by any country unilaterally is stamped legal and fair, which puts every country in Hamlet's shoes and makes it ask the all-important question – To be or not to be? The upcoming elections in the US, France and Germany have led to a slump in the 'green' momentum leaving a huge question mark over these countries' future policies on environmental and climate change as well as their role in the negotiations.

In short, the pathway of the negotiations has to be redrawn so that the international community does not lose sight of what could be in store for the humanity. It is not about the 'Doomsday' phobia; it is about raw practicality that could save millions of lives that are currently being endangered by faulty policies of the international set-up.

Notes

1. Anjali Jaiswal and Shravya Reddy, "The Greening of U.S.-Indian Relations: A Review of Cooperation between the United States and India on Climate

Change and Energy," *India Facts* (Washington, DC: Natural Resources Defence Council, January, 2011), pp. 1-2, see http://www.nrdc.org/international/india/files/usindiagreeningsummary.pdf, accessed on February 05, 2011.

2. EU-India Initiative on Clean Development and Climate Change, British High Commission, New Delhi, for more information, August 11, 2008, see http://ukinindia.fco.gov.uk/content/en/article/ClimateChange/5158547, accessed on October 08, 2011.
3. See, "China, India Sign Agreement of Co-Op on Climate Change," *People's Daily Online* (October 21, 2009), http://english.peopledaily.com.cn/90001/90776/90883/6790138.html, accessed on October 08, 2011.
4. See, Climate Avenue, "Clean Energy or Renewable Energy in China," March 17, 2011, http://www.climateavenue.com/China.en.renewable.htm, accessed on September 17, 2011.
5. Sharon LaFraniere, "Chinese Protestors Accuse Solar Panel Plant of Pollution," *The New York Times* (September 18, 2011), see http://www.nytimes.com/2011/09/19/world/asia/chinese-protesters-accuse-solar-panel-plant-of-pollution.html, accessed on September 18, 2011.
6. Jonathan Watts, "China Warns of 'Urgent Problems' Facing Three Gorges Dam," *The Guardian* (May 20, 2011), see http://www.guardian.co.uk/world/2011/may/20/three-gorges-dam-china-warning, accessed on September 18, 2011.
7. Cleo Paskal, *Global warring: How Environmental, Economic and Political Crises will Redraw the World Map* (Toronto, Ontario: Key Porter Books Limited, 2010), p. 152.
8. Interaction with Shyam Saran, Former Prime Minister's Envoy of Climate Change and Former Foreign Secretary, at Research and Information System for Developing Countries (RIS), on June 15, 2011.
9. Interaction with Niu Qingbao, Consul General of the People's Republic of China, at Manipal University, on October 14, 2011.
10. Interaction with Group Captain Manoj Kumar, Centre for Air Power Studies (CAPS), New Delhi, on September 15, 2011.
11. Cleo Paskal, "Why the West is Losing the Pacific to China, the Arab League, and Just about Everyone Else," *Huffington Post* (November 24, 2010), see http://www.huffingtonpost.com/cleo-paskal/why-the-west-is-losing-th_b_786668.html, accessed on September 18, 2011.
12. Bryan Walsh, "Black Carbon: An Overlooked Climate Factor," *Time Science* (November 13, 2009), see http://www.time.com/time/health/article/0,8599,1938379,00.html, accessed on September 13, 2011.
13. See, Press Trust of India, "Black Carbon can't be Linked to Global Warming," *Business Standard* (November 16, 2009), http://www.businessstandard.com/india/

news/black-carbon-cant-be-linked-to-global-warming-pachauri/23/37/78409/on, accessed on September 13, 2011.

14. Interaction with Group Captain Manoj Kumar, Centre for Air Power Studies (CAPS), New Delhi, on September 15, 2011.
15. Kentaro Tamura, "Technology Development and Transfer," in Ancha Srinivasan (ed.), *Asian Aspirations for Climate Regime Beyond 2012* (Haryana: Institute for Global Environmental Strategies, November, 2006), p. 55.
16. See, World Bank, "Growth in Global Carbon Market Pauses amid Uncertainty," Press Release no. 2011/514/SDN, http://web.worldbank.org/WBSITE/EXTERNAL/NEWS/0,,contentMDK:22928671~pagePK:64257043~piPK:437376~theSitePK:4607,00.html?cid=3001_2, accessed on September 20, 2011.
17. n. 15, p. 63.
18. See, "Financing Quarrels Mar UN Climate Talks in Panama," *Environment News Service* (October 7, 2011), http://www.ens-newswire.com/ens/oct2011/2011-10-07-02.html, accessed on October 08, 2011.
19. See, "Delay climate treaty until 2015?," *Terra Daily* (September 21, 2011), http://www.terradaily.com/reports/Delay_climate_treaty_until_2015_999.html, accessed on September 21, 2011.
20. Meinhard Doelle, "The Legacy of the Climate Talks in Copenhagen: Hopenhagen or Brokenhagen?" *Carbon & Climate Law Review* (Germany: Lexicon Publisher, 2010), vol. 4, no. 1, p. 96.
21. See, "An Introduction to the Kyoto Protocol Compliance Mechanism," vp, accessed on September 21, 2011.
22. Email correspondence with Dr. Matthew Paterson, University of Ottawa, dated August 06, 2011.
23. Damian Carrington and Allegra Stratton, "What Happened to the Conservatives' Green Agenda?" *The Guardian* (October 6, 2011), see http://www.guardian.co.uk/environment/2011/oct/06/david-cameron-green-agenda-fades?fb=native&CMP=FBCNETTXT9038, accessed on October 08, 2011.
24. Tom Young, "Bonn Climate Change Talks: What Next?" *business Green* (June 9, 2011), see http://www.businessgreen.com/bg/analysis/2077439/bonn-climate-change-talks, accessed on October 03, 2011.
25. See, "Japan, India Halt Move to Ban Nuclear in CDM: Source," *Point Carbon* (October 4, 2011) http://www.pointcarbon.com/news/1.1594322, accessed on October 08, 2011.
26. Navroz K. Dubash and Lavanya Rajamani, "Beyond Copenhagen: Next Steps," *Climate Policy* (London: Earthscan, 2010), issue 5, p. 596.
27. Email correspondence with Dr. Matthew Paterson, University of Ottawa, on August 06, 2011.

28. Interaction with Cleo Paskal, Associate Fellow, Chatham House, at Manipal University, on September 20, 2011.
29. Bryan Walsh, "Water: Lake Mead Is at Record Low Levels. Is the Southwest Drying Up?," *Times* (October 18, 2010), see http://ecocentric.blogs.time.com/2010/10/18/water-lake-mead-is-at-record-low-levels-is-the-southwest-drying-up/, accessed on October 19, 2011.
30. Fiorella Gardella, "The Meat Industry: Is it Worth it?," Seminar in Global Sustainability (Irvine: University of California, March, 1999), see http://www.dbc.uci.edu/sustain/global/sensem/MeatIndustry.html, accessed on September 25, 2011.
31. "Fight Global Warming by Going Vegetarian," PETA, for more information, see http://www.peta.org/issues/animals-used-for-food/global-warming.aspx, accessed on September 25, 2011
32. Interaction with Navdeep Suri, Joint Secretary, Public Diplomacy Division, Ministry of External Affairs, Government of India, at Manipal University, on October 07, 2011.
33. Interaction with Brajesh Mishra, former National Security Advisor and Principal Secretary to the former Prime minister, A. B. Vajpayee, at official residence, New Delhi, on July 07, 2011.

5

Environment and Energy: Roadblocks and Prospects for India

> "Winston Churchill aroused this nation in heroic fashion to save civilisation in World War Two. We have everything we need except political will, but political will is a renewable resource."
>
> — Al Gore, Former Vice-President of the US and Nobel Laureate

The drivers of India's climate change policy at the domestic and international levels have been touched upon in the second chapter. What one needs to ponder over is whether India should continue to follow the Western models of economic growth which is now slowly tilting towards low-carbon growth or should it create an indigenous model that would be in harmony with the Indian social, political, economic, environmental, psychological and cultural set-up. Experts say, "India has a lot to offer. In spite of all odds, and against much opposition, it is a more or less stable (and the less is not the fault of 'average' India), secular (but moderate, and with soul), people-oriented, growing economy where people still believe in democracy. Each of those components is rare, valuable, and in global short supply."[1] At the same time, it is quite obvious that there is a discernable difference between having the potential and utilising or harnessing it. The negotiations might break down from time to time but it would still continue as nation states could at least use the neo-liberal institutional rhetoric as a camouflage for fulfilling their 'realistic' goals. Therefore, what is being followed elsewhere in the world and what India is being persuaded to follow need not be successful in India. However, an idea of the international discourse on various adaptation and mitigation strategies is relevant.

The debate surrounding black carbon was reflected upon in the fourth chapter. Similar debates on various other issues have ensured that India's stand on climate change policy is diluted to such an extent that it has lost vision of its goals. One need to ask this question constantly – is it enough to have sustainable development in one's ethos? The development or environment debate that pervades across the policy-making mechanism in the country has depicted the two concepts to be antithetical to each other. The reality is quite different. All the developmental activities could be obliterated by unfathomable changes in environment. It took just one tsunami to trigger off the Fukushima disaster in Japan. It took one Hurricane Katrina to damage and shut innumerable oil wells and natural gas pipelines along the US Gulf Coast. If everyone concentrates on pin-pointing whether these events are set off by climate change or not, nobody could reach any durable solutions to protect these infrastructures that are symbols of development.

The international community has begun to emphasise on the argument that a breakthrough in the negotiations could occur only if there is a technological breakthrough. Unlike the Montreal Protocol that has been largely successful due to technological breakthrough, the Kyoto Protocol or any other climate change protocol could face several blockades due to a lack of technological replacements that the international community is in search of. Although there are thousands of technologies available that could address the GHGs including renewable energy technology, lack of consensus and political will would come in the way of the implementation of a robust international agreement.

The onus lies on India to find its way through the different gates of the climate change debate and reach its destination. In this context, this chapter would first focus on the salient features of the Interim Report of the Expert Group on Low Carbon Strategies for Inclusive Growth (2011) of the Planning Commission as it would serve as a reference point for India's future climate change policy. The chapter would reflect on the various measures to address climate change issues such as Renewable Energy Technology (RET), nuclear energy, carbon/CO_2 Capture and Storage (CCS), emissions trading, carbon tax, and so on. Attention would be given to possible adaptation measures that India could espouse. The attempt in the chapter would be to analyse the issues from a policy-perspective and not from a technical or scientific viewpoint.

Policy-Making in India: Words and Actions

What is 'in' the report?

Though the Interim Report leaves several questions unanswered, it is important to have a look at it for three reasons – first, it shows the pulse of the policy-making system; second, it allows the critics to evaluate the dangers of straight-jacketing the whole environmental change to just climate change, and third, it allows the critics to differentiate between the country's domestic policy and international policy. This report might have been prepared for the international audience to show India's commitment towards mitigation. Yet the Government needs to draw a line between its international policy and domestic policy for better understanding of the domestic audience. In fact, the report itself unusually justifies the loopholes in it. As already discussed in the second chapter, this report has outlined the 'Determined Efforts' and the 'Aggressive Efforts'. These efforts could be employed in various sectors that are known to be the highest contributors to carbon emissions in the country. Even on the energy efficiency front, the report has delineated two scenarios – efficiency gains through 'Determined Effort,' in which case 70 per cent of all appliances sold in 2020 would be highly efficient and the remaining would use current technologies; and efficiency gains through 'Aggressive Effort', in which case 65 per cent of all appliances sold in 2020 would be super efficient, 25 per cent would be highly efficient and the remaining would use current technologies.[2] The report also makes various recommendations on the technological and energy fronts such as super critical power plants to improve combustion efficiency and which could contribute up to 50 GW by 2050[3], the CCS, gas based power, hydro and renewable sources and nuclear power. In the transportation sector, the report has suggested five mitigation options—(1) improving the efficiency of road freight, (2) increase the share of train in freight transport, (3) promoting public and non-motorised transport, (4) improving fuel efficiency of vehicles, and (5) introducing cleaner fuels such as bio-diesel.[4] The emissions from the transport sector could be reduced only if the technological advancements are supplemented by responsible and optimal behaviour by the human beings. If people could start car pooling or use public transport or avoid travelling if some work can be done through telecommunication, emissions could be reduced to some extent. In the buildings sector, the report has thrown light on the importance of implementation of Energy Conservation Building Code and Green Building Rating System to reduce electricity consumption. In the industry sector, the report delves into four key areas – iron and steel, cement, oil and gas. Yet again, the suggested measures are more or less 'old wine in a new bottle'. In fact, in the case of the industry sector, the report has failed to come up with any solid recommendations. Very importantly, the report focuses on the institutional set up that would play an equally important role in facilitating the implementation of some of the measures suggested in the report. Some of them include constant technological and institutional innovation, self-regulation, implementation at and through all levels, participation of non-governmental organisations, coordination to reduce transaction costs, factoring of uncertainties in the context of climate change, pricing of fossil fuels, and so on.[5]

Future of Renewable Energy

Renewable energy technologies would be one of the best options for India to optimise its energy consumption and yet not degrade the environment to the extent that it is doing right now. One of the biggest advantages of renewable energy is that this form of energy could be deployed at the local levels on a small-scale or in a decentralised manner so that the consumption of energy is less and the distribution is easier and sustainable. One has to realise that all large-scale projects need not be stable and sustainable. The special report of the Inter-governmental Panel on Climate Change (IPCC) has come out with several findings about the potential of renewable energy. One of the most important points that it has raised is that 'the total global technical potential for renewable energy is substantially higher than global energy demand'.[6] It also contends that between 2010 and 2050, renewable energy could make global cumulative savings of CO_2 from about 220 to 560 Gt CO_2 compared to about 1,530 Gt cumulative fossil and industrial CO_2 emissions in the IEA World Energy Outlook 2009 Reference Scenario during the same period.[7] If the potential exists, India should exploit it before it misses the bus. At the same time, it has to overcome several challenges connected to the renewable energy technology. For example, the renewable energy units themselves could be affected by climate or environmental change. Large-scale hydropower projects in many parts of the world including India are currently under stress due to drastic variations in the amount of rainfall and glacial recession.

In India, the rivers in the South are monsoon-fed and therefore, highly vulnerable. Huge solar power plants could be flooded or hit by earthquakes and permanently destroyed if they are built in flood plains or seismic zones respectively. Therefore, they need to be designed in such a manner that they are protected from possible climatic variability and natural disasters. There are also cases in which such power plants have also been accused of polluting water bodies by releasing toxic substances into them such as in the case of solar power projects. If the solution to environmental degradation itself becomes a cause for it, the whole purpose of building a greener future is defeated. It is also imperative for India to make renewable energy economically competitive. Imposition of higher taxes on the supply and production of fossil fuels as well as subsidies on the production of renewable energy could play a positive role in the long run. The costs could come down further due to advancements in technology, proliferation of projects and greater demand. In the Interim report, one of the other major roadblocks in the mainstreaming of renewable energy technologies has been identified as their integration with the current and future energy systems. A few experts have even recommended the setting up of Market Based Instruments such as Green Pricing Schemes for renewable electricity with tradable renewable energy certificates to incentivise the use of renewable energy by the industries.[8]

Some of the interesting, optimistic or even astounding stories and proposals that could pave the way for a greener future include:

- Mahindra Reva Electric Vehicles Private Limited, [earlier a joint venture between Reva Electric Car Company started by Sudarshan Maini and Amerigon Electric Vehicle Technologies (AEVT) of the US] a company based in Bengaluru, is the pioneer of electronic vehicle technologies with zero emissions.[9]

- The Tatas are planning to bring the one-turbine-per-house project to the rural areas in India that are not connected to the national electricity grid. This is another example of how decentralised wind power units could make a huge difference by not using up vast expanses of land and other resources.[10]
- India is now the pioneer of 'low carbon aviation fuel with just half the carbon footprint of the standard fossil fuel alternative'. It has managed to produce this fuel in partnership with LanzaTech (of New Zealand) by a process in which the 'waste gases from industrial steel production are captured, fermented and chemically converted using Swedish biofuel technology'.[11] Civil aviation is one sector in which the world has so far not made any technological breakthrough in terms of reducing its emissions. This could indeed prove to be the first breakthrough and if this works, the carbon emissions could be reduced by a great degree.
- Desertec Industrial Initiative is presently building several solar power plants across North Africa. Thereafter, it plans to ship electricity up north via power lines under the Mediterranean Sea. More than 30 companies have now joined the initiative including Germany's Siemens, Swiss-Swedish group ABB, etc.[12]
- A Japanese company, Shimizu Corporation, intends to use robots to set up solar arrays (6,800 mile belt) on the moon and beam the energy back down to the earth. It could provide 13,000 TW of energy for the earth reportedly.[13]
- Winflex Wind Turbines of Israel is said to have developed an inflatable wind turbine made out of cloth which can be installed on roofs or on the ground and is less expensive than conventional wind turbines.[14]
- The world's first Gemasolar—a solar thermal plant with a major solar tower—has become operational near Seville in Southern Spain. It stores energy, often in molten salt. The speciality of this plant is that it produces electricity even at night. It is expected to produce 110 GWh/year—enough to power 25,000 homes in Andalucia region.[15]
- 'Kites flying under the sea' could be used to harness tidal energy. Minesto of the UK is developing 'Deep Green' technology which according to the company could generate enough electricity (530 GWh) to power all the

homes in a city of the size of Newcastle (population 189,863) by 2020.[16]

- According to the University of Wageningen microalgae can supply the whole of Europe with biodiesel which could be the best possible alternative for fossil fuels.[17]

India could learn a few lessons from these stories and try to emulate them as it is a storehouse of abundant renewable sources. It is also necessary to highlight some of the planned ventures of the country in this context. They are:

- An estimated 13,000 MW could be garnered from the existing solar energy infrastructure alone. The three-phased National Solar Mission aims to scale up the country's solar energy capabilities drastically in order to address the challenges of both energy security and climate change. 'The first three years would see a push to scale up research to drive down costs, spur domestic manufacturing, and fund solar lighting with micro-finance. The second phase would require installation of solar water heaters in office buildings and large residences. In the third phase, which India hopes will begin by 2017, solar energy is to achieve tariff parity with conventional power'.[18] In 2008, Clinton Foundation pledged to set up an US$ 8-10 billion 'Integrated Solar City' project in Gujarat with a generational capacity of 3,000 MW, which could go up to 5,000 MW in which case the costs would go up to US$ 15 billion. The cost of generation of solar power in India is expected to be lower than even a coal-fired plant in the next few years.[19] Talks are on to set up such projects in Andhra Pradesh and Rajasthan as well. Also, a string of corporates including Essar, Indiabulls, Reliance, ADAG, Tata Power, Suryachakra and Euro Group have expressed their willingness to collaborate with the Government. The Mukesh Ambani-controlled Reliance and Euro Solar have already entered into partnerships with the Government.
- GE Energy Financial Services and Greenko Group Plc have announced a venture to develop wind energy projects costing US$ 115 million. Greenko is developing a series of wind projects in Maharashtra, Andhra Pradesh, Karnataka and Rajasthan. "Once operational, a 500 MW

wind portfolio could generate enough renewable electricity to power 8,75,000 average Indian households and displace 7,00,000 tonnes per year of greenhouse gas emissions (GHGs)."[20]

- Gujarat could host Asia's first commercial-scale tidal power station by 2013. The 50 MW tidal farm is expected to cost US$ 150 million. It could be further expanded to deliver 200 MW.[21] Before this, the country had also planned to set up a minor tidal power generation project with an installed capacity of 3.75 MW at Durgaduani Creek of the Sunderbans; it was cleared in 2008. It is a joint venture in which both the Government of India as well as the State Government of West Bengal has invested.[22]
- Biomass remains the principal energy source in rural and traditional sectors and it contributes 32 per cent of the total primary energy use. The Government's strategy has centred on 'improving efficiency of traditional technologies, enhancing supply of biomass, introducing modern biomass technologies to provide reliable energy services at competitive prices and establishing institutional support'. A programme whose inception was facilitated by the Department of Non-Conventional Energy sources was later on consolidated by the Ministry of Non-conventional Energy Sources' renewable energy-centric strategy. Analysts such as P. R. Shukla believe that 'biomass electricity penetration can reach 35 GW in 2035'.[23]

Peaceful Use of Nuclear Energy

Nuclear renaissance in India is a long way away. The debate between the pro-nuclear lobby and the anti-nuclear lobby is a longstanding one and a glimpse of it was presented in the second chapter. The growing demand for energy and the urgent need for the international community to address environmental issues such as climate change have to be addressed in tandem by nation states. India is of the firm belief that nuclear energy is the best way forward to meet those twin goals. However, the recent mishap at the Fukushima Daiichi nuclear power plant caused by the earthquake and the tsunami in Japan has raised a few concerns among the public not only in India but across the globe. The Jaitapur plant in Maharashtra has already come under scrutiny for a number of reasons. The Evolutionary Power Reactors (EPR) produced by the French company Areva have not been commissioned anywhere in the world plus they are more expensive compared to other reactors. Certain political parties and civil society groups have been questioning the safety and reliability of these reactors since they are the biggest nuclear reactors in the world. These fears have been allayed to a certain extent by the Government as well as the Atomic Energy Commission that have upheld their view that the EPRs are absolutely safe. Moreover, they also affirm that Jaitapur falls under the Seismic Zone 3 category, thus it cannot be considered a highly earthquake-prone area. At the same time, reports suggest that the area has experienced as many as 92 earthquakes between 1985 and 2005.[24] It is 75 ft above sea-level and the height of tsunami waves across the world has been between 20 and 30 ft.[25] In fact, India's reactors are regarded as some of the safest ones in the world and until now no major mishap or accident has occurred. For example, Kalpakkam nuclear reactors remained unaffected after the 2004 tsunami as they shut down automatically, although water had seeped into their drainage systems. On the contrary, reports brought out by Greenpeace and Tehelka disclose a diametrically opposite story which is filled with incidents such as 'leaks, fires and structural damages'[26] and of course 'cover-ups'[27] in which workers' lives as well as that of the people living in the surrounding areas have been endangered due to their exposure to radioactivity.

Recent reports have also revealed that not only has the EPR construction in France been delayed due to large number of accidents including two fatalities, its price tag has also climbed greatly,[28] raising more doubts in the minds of the Indian public. Nuclear power undoubtedly is an inherently unstable form of energy. At the same time, one could say with absolute conviction that India has not witnessed an accident of the scale of Three Mile Island (1979), Chernobyl (1986) and Fukushima (2011). Another roadblock that the Government has been confronting and is expected to confront in future in terms of implementation of the Nuclear Deal would be acquisition of lands, especially in heavily populated areas. The protests against the Kudankulam Nuclear Power Plant have also raised questions over the future of nuclear energy in India. The findings of the Home Ministry that these protests had received funds to the tune of Rs. 55 crore from foreign sources[29] reinforces the belief that the there are vested interests involved in stalling India's nuclear programme.

The counter-claims have also revolved around certain perceptions associated with nuclear power. If nuclear power can be delinked with possible spread of nuclear weapons, nuclear accident and nuclear waste disposal, then nuclear power could be a solution for tackling the sheer shortage of electricity in the country. Anil Kakodkar, former chairman of the Atomic Energy Commission, believes that India should not roll back its nuclear programme as 'nuclear renaissance' is the only solution to India's energy security. According to him, since the additional energy required in India is high, even higher than the additional energy required in China and since development requires energy, India has to dispose of the 'catastrophe syndrome' and take the nuclear leap. In North America and Western Europe, the nuclear power generation equals roughly 15,000 KW per capita; 10,000 KW per capita in the OECD countries and the world average is approximately 2,800 KW per capita. In India, the figure stands at a meagre 700 KW per capita. India has to go a long way to reach the global average as well as at least be comparable to the OECD countries. Since India's population is not yet stable and could grow by 20-40 per cent by 2050, the energy required would also increase by 20 times. Thus, even if India manages to touch 5,000 KW per capita, it would still require 10 times more energy and if it reaches 2,500 KW per capita, it would face an alarming 50 per cent shortage by 2050. Fifty years down the line, the oil deposits might not exist while prospects for uranium supplies are tremendous. Fructification of full-scale utilisation of the available thorium reserves as well as fusion would make sure that no other source of energy would be able to match up to nuclear energy in terms of power generation capacity. As far as other renewables are concerned, hydropower is not only capital intensive but also it leads to disturbance in the eco-system, displacement of people and emission of methane from decaying bio-energy in reservoirs.[30] Solar power is constrained by the efficiency of the commercial cells, and wind power may not be practical in all places. The risks of CCS would be discussed later in the chapter. In such a scenario, he insisted that nuclear power was more of a necessity, a matter of no choice. Kakodkar also stressed on the need for media to play a constructive role in creating public awareness about nuclear energy rather than harping on its shortcomings or hazards.[31]

Nuclear renaissance is being widely debated across the world. The debate took shape in the US especially in the aftermath of the Three Mile Island and the Chernobyl disasters. The US did not build a single nuclear reactor after the Three Mile Island accident. The debate gained momentum only in the twenty-first century when concerns over rising fuel prices and climate change gained salience. Countries like France, Finland, Japan and South Korea have made significant progress in the nuclear power industry revival. Countries such as China, Israel, the UK, Switzerland, Thailand, Malaysia and the Philippines had started to review their nuclear programmes to explore this option. However, the Fukushima crisis has weakened the movement towards achieving nuclear renaissance across the globe. This has clearly forced the energy-consuming countries to re-evaluate the safety of their existing reactors and convey apprehension over the pace and magnitude of planned expansions. For example, Germany shut down seven of its oldest reactors immediately after the Fukushima crisis erupted. It has also decided to phase out nuclear power completely by 2022. Several other countries such as Switzerland and the UK have also put their nuclear expansion plans on hold. India on the other hand, has refused to halt the nuclear industry revival process and on the occasion of signing the India-Kazakhstan nuclear agreement, Manmohan Singh pointed out, "In our view, nuclear energy is an important component of our energy mix in our national strategy to meet our future energy needs. In the light of the tragedy in Japan, we have embarked on a comprehensive review of the security and safeguard procedures at our nuclear plants. We look forward to our continuing cooperation with Kazakhstan in the peaceful uses of nuclear energy."[32]

It is estimated that today's nuclear reactors across the world are preventing emissions of 2.5 billion tonnes per year of CO_2 – which is about half the CO_2 emitted by the world's motor vehicles. Nuclear power has the capacity to generate large amounts of electricity unlike fossil fuels.

Nuclear power plants do not emit GHG and the waste produced is very little compared to the amount of electricity produced. Environmentalists have been divided on the question of nuclear power as some believe that the radioactive waste could lead to environmental and health hazards. This could be counteracted by the claim that since the inception of the nuclear industry, waste management has been of top priority for the industry unlike the other industries that have been haphazard in terms of waste management causing considerable amount of environmental damage. Dr. Manpreet Sethi stresses that after the capital investment, the cost of generation of nuclear power per unit energy is lower than other sources. According to her, the safety and security of the nuclear power plants are intact with the building of new designs that could deter air crashes or detonation of conventional explosives besides the fact that every nuclear plant has several layers of security other than the double containment structures. The need of the hour is to beef up security and intelligence in order to prevent untoward deliberate accidents. The Indian establishment has to be more proactive in terms of creating awareness about this form of energy and dispelling fears and superstitions. Nuclear energy is a mature technology in comparison to several renewable energy technologies and thus one should keep this option open for all practical reasons.[33]

The Institute for Global Environmental Strategies has discovered a sea-change in the attitude of the Japanese public towards nuclear power. Three-fourth of nuclear power plants was shut down after the Fukushima accident. Only 3 per cent of population wants nuclear power to continue. 34 per cent of the population wants all nuclear power plants to close in the next two years; only 34 per cent of the population agrees to the continuance of functioning of plants till the end of their operating lives.[34] The process of setting timelines of phasing out nuclear power is already underway with the election of the new Prime Minister—Yoshihiko Nada (sixth Prime Minister in five years)—after Naoto Kan was forced to resign due to increasing public pressure on account of his poor management of the Fukushima disaster. German Chancellor Angela Merkel has declared, "We can be the first major industrialised country that achieves the transition to renewable energy with all the opportunities—for exports, development, technology, jobs—it carries with it."[35] In such a scenario, India should set short-term, medium-term and long-term goals for itself to guarantee energy security for a long period of time. The delay in supply of the EPRs and the introduction of new Nuclear Suppliers' Group (NSG) rules for enrichment (ENR) technology have put a question mark over the possibility of meeting short-term and medium-term targets in terms of nuclear power. Therefore, coal, oil and natural gas would still be the major sources of energy at least till 2020. The period after that might see a slight shift towards nuclear energy as envisaged by the Indian authorities with fossil fuels still fulfilling more than half of the energy needs. India has to think in terms of long-term targets and it is highly possible that countries across the globe especially the industrialised ones would have already made their transition to renewable energy successfully by 2050. Nuclear energy could be treated as a bridge power in India's context. The only question is whether the country should wait for the whole world to master its R&D in renewable technology and then import these technologies for its domestic market or should the country get its R&D off the ground in a full-fledged manner like the Western countries before having to depend on foreign technologies. The decision of India and the US to jointly contribute US$ 50 million for 'joint, collaborative research in three critical fields: solar energy, second-generation biofuels, and energy efficiency in buildings'[36] is a welcome step. The key for India is to create a diversified energy portfolio and not put all its eggs into one basket.

Capture Carbon and Save the World?

'CO_2 Capture and Storage (CCS) is a process consisting of the separation of CO_2 from industrial and energy-related sources, transport to a storage location and long-term isolation from the atmosphere'.[37] While discussing about the CCS, one also has to analyse the capability of the available technology to capture CO_2, the amount of CO_2 emissions resulting from the energy required for capture, transport and storage as well as the possible leaks during transport. The three different CO_2 capture technologies are post-combustion (extracts CO_2 from the flue gas following combustion of fossil fuels or biomass), pre-combustion (separates CO_2 from the fuel by combining the fuel with air or steam to produce hydrogen and a separate

CO_2 stream that could be stored) and oxyfuel (uses oxygen instead of air for combustion and produces a flue gas that is mostly CO_2 and water, which are easily separable). The transportation to specific storage sites could be carried out either through pipelines or portable transport facilities such as trains, ships or trucks. Storage is usually done in geological formations such as deep saline aquifers or even depleted oil and gas reservoirs.[38]

India has already stepped into this arena through the establishment of the Indian CO_2 Sequestration Applied Research Network in 2007. It has also been actively participating in the Carbon Sequestration Leadership Forum and the International Partnership for a Hydrogen Economy. India's enthusiastic partnership with the US in this field (US FutureGen project and the US Big Sky CCS) has steered its efforts to deploy this technology in the country. The CO_2 storage potential estimates reveal that it is in the range of 500-1000 Gt of CO_2, including 'on-shore and off-shore deep saline formations (300-400 Gt), basalt formation traps (200-400 Gt), unmineable coal seams (5 Gt), and depleted oil and gas reservoirs (5-10 Gt)'.[39]

It has been found that the aquifer storage potential could be mostly tapped in Gujarat, Rajasthan and Assam. These aquifers are however located far away from the point sources that increases transport expenses. Offshore and onshore oil and gas fields have also been found to be insufficient in terms of storing emissions 'from even a medium-sized coal-fired power plant'.[40] The Indian establishment has thus been non-committal about the CCS technologies because of several factors—first, these technologies entail high expenditure as the cost of energy production goes up. The costs of the CCS start right from capture to transportation to storage and most importantly monitoring; second, the technologies are either not available at equitable terms or they are not foolproof. The US, the EU and Australia are home to a few pilot projects of the CCS. Due to imperfect capture process or storage, CO_2 can be released into the atmosphere, and third, inland storage sites need not necessarily be safe as they could also be seismically active which in turn could lead to leakage of CO_2. As Shyam Saran points out, "We should do what is good for us rather than blindly following the low carbon growth strategies being put forth by the West. This would only mean that we have accepted our permanent inferiority."[41] The larger question of

'Who will pay'? applies to this aspect as well. CCS is not yet part of the Clean Development Mechanism (CDM) activities. Even when it becomes part of the CDM, technology transfer will remain a bone of contention as resolution of the Intellectual Property Rights by the international community is a long way away. India has also expressed its apprehension with respect to the compatibility of the mechanism with the sink UNFCCC definitions. Since the CCS captures CO_2 from the point sources and not the atmosphere, it cannot be qualified as 'sink'. However, it could also not be considered emission reduction since the emissions in reality do not reduce. The only way it could be endorsed by the UNFCCC would be by counting the captured CO_2 as subtracted CO_2 emissions from the atmosphere. The difficulty in terms of setting the project boundaries would be contingent on identification of the source and sink of CO_2 since they could even be in two different countries belonging to different groups (Annex I and non-Annex I countries). The baseline and methodology setting would therefore prove to be a Herculean task as the study of the geological formations and the potential risks requires plenty of time and energy; its accuracy would always come under the scanner too. Moreover, the CCS technologies can succeed only if price of carbon is sufficiently high. For these reasons, India's approach to the CCS should be cautious. It is true that emissions cannot be curtailed by one or two methods and the world has to embrace a multi-pronged approach concentrating on different methods. The key to the success of the CCS would be a technological breakthrough.

Trade Dominates the Discourse

The Emissions Trading Scheme (ETS) is another important means of mitigation that is incentive-based. It has been found to be successful in Annex I countries and could now be imbibed by non-Annex I countries. India, China, Mexico and other developing countries have taken the first step to introduce this system. India's proposal of 'energy saving certificate' instead of 'emissions reduction certificate' has already been discussed in the second chapter. According to a study conducted by Prabhat Upadhyaya, first, ETS can work in India only if the entities covered in the ETS and the CDM are different since both systems 'operate with completely different principles towards the same objective: emissions reduction'. Second, the overlapping of the Perform Achieve Trade (PAT) scheme with the ETS is inevitable in a country like India. This is primarily because of the fact that usually an ETS covers energy-intensive industries as seen in the case of the European Union ETS, but in India's case, they are already covered by the PAT scheme which implies that such industries could enjoy the benefits of both systems through the same set of activities. As part of the NAMAs, India could easily establish the ETS and if they are credited, then they could even be utilised to meet Annex I countries' targets or non-Annex I countries' domestic voluntary targets.[42] In order to simplify the process and to make sure that the mandate of the ETS is fulfilled, the country should have precise and consistent GHG emissions data for a long period of time. In this context, as discussed in the second chapter, the Government has to ensure that its involvement should be restricted to deciding the baseline, the number of allowances, the compliance mechanism and so on. It should not interfere in the procedure of allocating permits to the industries. Once the standards are set, the industrial complex should be allowed to take it forward after signing a Memorandum of Understanding (MoU) with the Government.

According to reliable sources, the Government plans to do otherwise. However, India could also follow the upstream system in which trading of emissions occurs at the level of fuel providers instead of the downstream system in which trading occurs at the level of manufacturing industries that could also be covered under the ambit of the PAT scheme. Also, the Government has to choose between providing allowances for free or auctioning them. In the long-term, the Government might also have to think about linking the domestic market with the regional/global market in the post-Kyoto scenario. With the decision of the Government of India to publish the emissions inventory on a two-yearly basis, there is a need to expand the coverage of the inventory to each and every sector, and ultimately to each and every entity. For this purpose, the Government would either have to introduce stricter regulations by which the industrial complex would be forced to declare their emissions and individual environmental impact assessments. This could be achieved by strengthening the existing Carbon

Disclosure Project and implementing the National Environment Protection Authority (NEPA).

The feasibility of such a system in the domestic scenario is again debatable. The political viability of an ETS in the Indian context is debatable due to the fact that capping of emissions would still be stonewalled by conservative policy-makers and business groups as it could result in spurt in the costs of end-products leading to higher inflation. Industries could also attempt to divert financial resources towards procuring capital-intensive technology leading to retrenchment of labour. Linking of the domestic ETS with the global ETS could also face a series of blockades. For example, as Matthew Paterson states, "Linkage to the regional market is possible only if India adopts fixed emissions reductions targets which seems a distant possibility as of now. It would also be difficult to link the domestic ETS to an offset system like the CDM."[43] Its political feasibility is again questionable as India would not be able to join the international emissions trading regime unless it is converted from the Grandfathered Emissions Allocation Scheme (GEAS) into the Equal Per Capita Emissions Allocation Scheme (EPCEAS). Only in such a sceanrio, India would have sufficient allocations to sell.

In the GEAS, emission permits are allocated on the basis of the aggregate emissions level of a predetermined year while in the EPCEAS, aggregate emissions entitlements for India for all the years are fixed at the level arrived at by multiplying the average global per capita emission with India's population for a benchmark year.[44] Developed countries would not be ready to do this as this would force them to buy emissions permits from developing countries which have low per capita emissions as compared to them. India, like any other developing country, would also not agree to the GEAS as it could have detrimental effects on its growth objectives. Therefore, India should wait and see how much of a success the PAT scheme would be and build the ETS on a similar foundation.

Adapting to the Changes

Currently, on the global adaptation index, India is ranked 36th, well below countries such as El Salvador, Tonga, Botswana, Kazakhstan, and so on.[45] The rating itself could be debated but the discussion could be moved forward by taking up the issue at the domestic level for strengthening India's adaptation policy. The sectors of concern are agriculture, water resources, health and sanitation, forests and extreme weather events. While framing policies for climate change adaptation, India needs to clearly lay equal emphasis on sustainable development and synchronise the two. Indian agriculture is heavily dependent on rainfall. With changing patterns of rainfall, run-off, soil moisture storage and increasing incidents of droughts and floods in different parts of the country, the Government has to adopt strategies that would make crops drought-resistant and flood-resistant. Two very important steps in this direction would be—first, reversion to the traditional ways of farming that was replaced by the Green Revolution to improve our food security, which would not only mean that resistance of the crops would increase but also quality of the soil would not be affected, and second, irrigation facilities have to be complemented by rain-water harvesting schemes and possibly ground water recharge systems which could be effective during excessive rainfall.

The use of biotechnology for designing both drought-proof and flood-proof crops is well-known; however, scientific research also has to complement the natural make-up of crops. The development of flood-resistant rice by the International Rice Research Institute along with drought-resistant genetically modified crops by Performance Plants Incorporated in Kingston, Ontario has not gathered much momentum due to stiff opposition from the traditionalists and environmentalists. Genetically modified food crops, which were touted as a breakthrough in terms of fighting pests and insects once, are now in the news for all the wrong reasons. Saline water, which is harmful for crops, would have to be desalinated using reverse osmosis plants. For this, such plants have to be made available at cheaper costs across the spectrum. These plants presently require large amount of energy and thus the need for optimisation of energy consumption as well as exploration of sustainable sources of energy to run them arises. Their use is quite prominent in cities such as Chennai, where the sea is the predominant source of potable water. The stress on water resources is expected to increase in the coming years. Community-based irrigation systems and water conservation would play a crucial role in adaptation. Excessive run-offs have to be prevented by diverting flood waters to storage bodies including small-scale dams. Heavy dependency of various industries including agriculture and iron and steel has to be reduced as well. Coastal areas have to be protected by building dikes, high tide lines, or other protective shields. The Government also has to make sure that the mangroves along the coastal areas are not destroyed.

Risk financing is an important element of adaptation and its importance in agriculture is well-acknowledged with the recent spate of farmer suicides. Welfare schemes are very important at a time when a large amount of activities that drive a country's economy could be affected by various changes in the environment. For example, the energy infrastructure of the country needs to be safeguarded from all eventualities that could affect India's energy security. The Jamnagar petroleum refinery—one of the biggest refineries in the world—was hit by a major cyclone in 1998 that created major havoc in the region; this despite the fact that its owners, Reliance had claimed that the location was safe as its deep harbours would protect the complex from storms. Jamnagar is one of the most flood-prone districts in India. As famously proclaimed, 'nature creates hazards, but human action creates disasters'.[46] Therefore, it is imperative for the Government of India to make sure that as an adaptation strategy, it has to prevent people from moving into natural disaster prone areas as much as possible besides the fact that critical as well as hazardous infrastructure should be designed in such a manner that they are either disaster-resistant and easily manageable/ rectifiable in case of a disaster or they should be built in safer locations. The role of Environmental Impact Assessment becomes crucial. Therefore, if an analysis of the costs and benefits is carried out, one would come to the understanding that prevention is better than cure. Between 1986 and 2006, India borrowed US$ 8257 million in the form of 43 loans for the purpose of disaster relief and rehabilitation. It was well ahead of China (32), Bangladesh (28) and Brazil (27). "The World Bank estimates that the material cost of disaster damage rose from barely US$ 40 billion in the 1950s to US$ 652 billion in the 1990s. The number of major disasters increased from 100 in 1975 to over 400 in 2005. Nature creates hazards, but human action creates disasters."[47] The reason why India's damage costs skyrocketed in comparison to countries such as China and Brazil that are larger in area is due to the population pressure. As the population keeps rising, risks increase and thus, the challenges ahead of the Government in terms of securing human security are immense. Moreover, the UN estimated that natural disasters caused economic damage of US$ 109 billion in 2010, three times more than in 2009.[48] Hence, adaptation is a huge challenge. It is very clear from these figures that the benefits of adaptation and mitigation far outweigh the costs as years pass.

Time Not on Our Side

India's adaptation and mitigation strategies are extremely insufficient at this point in time. Though options are galore, India has been spending more time on research rather than taking timely action. This is the understanding that has been inferred from the interactions with several experts and policy-makers in the field. In this process, it has also been neglecting its own in-built capacity that it has developed over the past few centuries and which could be mainstreamed. India has huge potential in the sphere of RET. Therefore, while investing huge financial resources in nuclear energy, it could lay equal emphasis on RET. Dr. Kirit Parikh speaks about the malicious IPR regimes that have made sure that there would be no balance in technology trade. India certainly would not like to go down the path of technological dependence; self-reliance is the key.[49] India should tread the path of emissions trading carefully so that corruption does not hijack such a system, while it could continue its research on the CCS for the time being. It is also high time that the country's administrators find ways and means of storing the excess electricity that is supposedly being produced in some of the States as mentioned in the second chapter. Another expert rightly says, "It is about preparedness. India could gain moral authority if domestic actions are in place."[50] Therefore, if India wishes to assert itself internationally, it has to go prove itself domestically.

What the climate change negotiations have done is to strengthen the existing poles in the world and even create new ones. The securitisation of the debate 'encourages a conflictual rather than cooperative and nationalist rather than internationalist stance'.[51] They have reverted the debate from the industrialised versus developing countries to man versus nature. In fact, the number of studies which hold solar changes and changes in the Earth's orbit responsible for climate change are on the rise. If Al Gore lashes out at the climate sceptics by comparing them to those who argued that smoking was not injurious to human health in the 1960s, there is a Rick Perry in the same country, who finds climate science a hogwash and highly misleading. The reality is that both of them have meddled with climate science by following either a reductionist approach (as in the case of the former) or a confrontationist approach (as in the case of the latter). By reducing the debate to the consumption of fossil fuels, Gore has lost a big opportunity to consolidate the public opinion against some of the evident changes in climate and environment and in turn, promote green technology – one of his biggest investments. Rick Perry does not even consider himself part of the debate. The end result is that even if these changes are natural, no country wants to take any steps to go any further in terms of adaptation or building resilience. Everybody is waiting for a low-cost technological breakthrough that is easily available.

It is already clear that one or two technologies cannot solve the problem. Instead of trying to make the CCS work so that emissions would not be deterred and 'unsustainable' development could go on, the international community should begin to plan its development in such a manner that its consumption of resources is less and the infrastructure is durable. Climate-proof infrastructure is a huge necessity in every single country. In this endeavour, the insurance sector, the planners and the military forces of a particular country need to be taken into confidence according to Cleo Paskal. The insurance sector could always put pressure on different entities to build only climate-proof structures which is a long-term solution. For example, Turkey has instituted mandatory national insurance in quake-prone areas. This would be extremely difficult to implement in several parts of India since the poor would obviously not be able to afford it. However, urban areas such as Mumbai and Delhi which are flood and earthquake prone respectively could have such mechanisms. Urban and rural planners could also help erect climate-proof, energy-efficient structures besides designing proper drainage systems, waterways, waste treatment strategies, and so on. The role of the military in protecting physical infrastructure and human life in times of natural calamities or disasters is well-known. Their first-hand experiences could also help in creating awareness and building pressure on the system to implement long-lasting solutions.[52]

Mindsets have to change in order to change the perception that adaptation and mitigation run opposite to each other with trade-offs between them. The reality is quite different. Without adaptation, the mitigation efforts would not sustain as they could be washed away by unfathomable changes in nature. Similarly, development is not possible without environment. Environment is our infrastructure. Therefore, 'adaptation' and 'mitigation' should continue to be used as catchwords at the appropriate forums for debate and discussion while the ground-level policies should follow sound organisational, institutional and communitarian principles.

Notes

1. Email correspondence with Cleo Paskal, Associate Fellow, Chatham House, dated October 10, 2011.
2. "Interim Report of the Expert Group on Low Carbon Strategies for Inclusive Growth," *Planning Commission, Government of India* (May, 2011), p. 32, see http://planningcommission.nic.in/reports/genrep/Inter_Exp.pdf, accessed on June 20, 2011.
3. Ibid., p. 33.
4. Ibid., pp. 46-57.
5. Ibid., pp. 110-111.
6. "IPCC Special Report on Renewable Energy Sources and Climate Change Mitigation," Summary for Policymakers approved at the 11th Session of Working Group III of the IPCC at Aby Dhabi, U. A. E., June 9, 2011, p. 7, see http://www.ipcc.ch/news_and_events/docs/ipcc33/SRREN_FD_SPM_final.pdf, accessed on September 12, 2011.
7. Ibid., p. 20.
8. P. R. Shukla, Amit Garg, Debyani Ghosh and P. Venkata Ramana, "Renewable Energy Technologies: Mitigation Potential and Operational Strategies," in P. R. Shukla, Subodh K. Sharma and Venkata Ramana (eds.), *Climate Change and India: Issues, Concerns and Opportunities* (New Delhi: Tata McGraw-Hill Publishing Company Limited, 2002), p. 267.
9. Ashok Parthasarathi, "Renewable Energy: A Major Instrument for Sustainable Living in the 21st Century," in Air Vice Marshall Kapil Kak AVSM, VSM (Retd) (ed.), *Comprehensive Security for an Emerging India* (New Delhi: Kalpana Shukla Publishers, 2010), p. 299.

10. See, "Rural India could be Powered by Small Roof-Mounted Wind Turbines," *Greener Tomorrow* (August 18, 2010), http://www.tomorrowisgreener.com/rural-india-could-be-powered-by-small-roof-mounted-wind-turbines/, accessed on October 08, 2011.
11. "Low Carbon Aviation Fuel Developed," *The Hindu* (Mangalore: October 12, 2011), p. 20.
12.. See, "What would it Cost to Power Europe with Solar Energy from the Sahara Desert?," *Greener Tomorrow* (January 3, 2011), http://www.tomorrowisgreener.com/what-would-it-cost-to-power-europe-with-solar-energy-from-the-sahara-desert/, accessed on October 08, 2011.
13. See, "Japanese Company Dreams of Building 6,800 mile Solar Array on the Moon," *Greener Tomorrow* (August 19, 2011), http://www.tomorrowisgreener.com/japanese-company-dreams-of-building-6800-mile-solar-array-on-the-moon/, accessed on October 08, 2011.
14. See, "The Winflex Inflatable Wind Turbine, ROI of Two Years," *Greener Tomorrow* (August 8, 2011), http://www.tomorrowisgreener.com/the-winflex-inflatable-wind-turbine-roi-of-two-years/, accessed on October 08, 2011.
15. See, "Making Light Work of it: The world's First Solar Power Station that Generates Electricity At NIGHT," *Mail Online* (June 3, 2011), http://www.dailymail.co.uk/sciencetech/article-1393879/Gemasolar-Power-Plant-The-worlds-solar-power-station-generates-electricity-NIGHT.html, accessed on October 8, 2011.
16. See, "Underwater Kites: Minesto's 'Deep Green' Wave Energy Project Scores Prototype Funding," *Huffington Post* (March 24, 2011), http://www.huffingtonpost.com/2011/03/24/underwater-kites-minesto-deep-green_n_839995.html, accessed on October 8, 2011.
17. See, "Biodiesel from Algae to Supply the Whole of Europe!" *Greener Tomorrow* (October 7, 2010), http://www.tomorrowisgreener.com/biodiesel-from-algae-to-supply-the-whole-of-europe/, accessed on October 08, 2011.
18. Jawaharlal Nehru National Solar Mission: Towards Building SOLAR INDIA, see for more information, http://mnre.gov.in/pdf/mission-document-JNNSM.pdf, accessed on February 03, 2011.
19. Maulik Pathak, "Solar Power in India will Cost Less than Coal Energy in Five Years," *Business Standard* (September 13, 2009), see http://www.business-standard.com/india/news/solar-power-in-india-will-cost-less-than-coal-energy-in-five-years/369928/, accessed on October 8, 2011.
20. See, "GE, Greenko Announce USD 115 mn Wind Energy Venture in India," *The Economic Times* (October 10, 2011), http://articles.economictimes.indiatimes.com/2011-10-10/news/30263347_1_wind-projects-wind-energy-wind-turbines, accessed on October 11, 2011.

21. Richard Black, "India Plans Asian Tidal Power First," *BBC* (January 18, 2011), see http://www.bbc.co.uk/news/science-environment-12215065, accessed on October 08, 2011.
22. Ministry of New and Renewable Energy, see for more information, http://www.mnre.gov.in/prog-tidal.htm, accessed on October 08, 2011.
23. P. R. Shukla, "Biomass Energy in India: Policies and Prospects," paper presented at the workshop on *Biomass Energy: Key Issues and Priority Needs* organised by International Energy Agency (IEA), Paris, February 3-5, 2008, p. 1, see for more information, http://www.decisioncraft.com/energy/papers/ecc/re/biomass/bpi.pdf, accessed on February 17, 2011.
24. Viju B, "20 years, 92 quakes: Ground trembles beneath Jaitapur's feet," *The Times of India* (March 16, 2011), see http://articles.timesofindia.indiatimes.com/2011-03-16/india/28699015_1_jaitapur-nuclear-power-plant-earthquake, accessed on March 21, 2011.
25. See, Nuclear Power Corporation of India Ltd, "Jaitapur EPR Reactor," http://www.npcil.nic.in/pdf/JAITAPUR_EPR_REACTOR_article.pdf, accessed on March 21, 2011.
26. See, Greenpeace, "Accidents at Nuclear Power Plants in India," http://www.greenpeace.org/india/en/What-We-Do/Nuclear-Unsafe/Safety/Nuclear-accidents/Nuclear-accidents-in-India/Accidents-at-nuclear-power-plants/, accessed on September 17, 2011.
27. Kunal Majumdar, "Accident Sites - Radiation, Cancer, Blindness, Tardiness, Cover-Ups: The Lessons from the Kalpakkam Nuclear Facility," *Tehelka Magazine* (New Delhi: September 11, 2010), v. 7, issue 36, http://www.tehelka.com/story_main46.asp?filename=Ne110910Accident_Sites.asp, accessed on September 17, 2011.
28. Vaiju Naravane, "Post-Fukushima, France Breaks Silence on Nuclear Safety," *The Hindu* (Mangalore: October 11, 2011), p. 15.
29. "Kudankulam Protestors Received Foreign Funds: Minister," *The Hindu* (January 25, 2012), see http://www.thehindu.com/news/national/article2831270.ece, accessed on January 25, 2012.
30. Manpreet Sethi, "Balancing Increasing Energy Demand with Environmental Concerns – A Case for Nuclear Power," *Nuclear Power* (New Delhi: Centre for Strategic and International Studies, June 1, 2004), vol. 1, no. 17, p. 1.
31. Interaction with Anil Kakodkar, at Manipal University, on March 26, 2011.
32. See, "India, Kazakhstan Ink Civil Nuclear Cooperation Deal," *The Times of India* (New Delhi: April 17, 2011), http://articles.timesofindia.indiatimes.com/2011-04-17/india/29427543_1_civil-nuclear-agreement-nuclear-energy-highly-prospective-region, accessed on April 10, 2011.
33. Interaction with Dr. Manpreet Sethi, Senior Fellow, Centre for Air Power Studies (CAPS), New Delhi at CAPS, on June 08, 2011.

34. Interaction with Group Captain Manoj Kumar, Senior Fellow, Centre for Air Power Studies (CAPS), New Delhi, on September 11, 2011.
35. See, "Nuclear Phase-Out can make Germany Trailblazer – Merkel," *BBC* (May 30, 2011), http://www.bbc.co.uk/news/world-europe-13597627, accessed on September 17, 2011.
36. Shravya Reddy, "Can Indian and U. S. Ingenuity Lead the world to Clean Energy Solutions?," *Project Groundswell* (May 25, 2011), see http://projectgroundswell.com/2011/05/25/can-indian-and-u-s-ingenuity-lead-the-world-to-clean-energy-solutions/, accessed on October 2, 2011.
37. See, "Carbon Dioxide Capture and Storage," Summary for Policymakers, A Special Report of Working Group III of the IPCC, http://www.ipcc.ch/pdf/special-reports/srccs/srccs_summaryforpolicymakers.pdf, accessed on September 05, 2011.
38. Peter Folger, "Carbon Capture and Sequestration (CCS)," *CRS Report for Congress* (USA: June 19, 2009), p. 6-8, see http://www.fas.org/sgp/crs/misc/RL33801.pdf, accessed on September 05, 2011.
39. Prateek Bumb and Rituraj, "Carbon Dioxide Capture and Storage (CCS) in Geological Formations as Clean Development Mechanism (CDM) Projects Activities (SBSTA)," p. 4, see http://cdm.unfccc.int/about/ccs/docs/CCS_geo.pdf, accessed on September 05, 2011.
40. Ibid., p. 6.
41. Interaction with Shyam Saran, Former Prime Minister's Envoy of Climate Change and Former Foreign Secretary, at Research and Information System for Developing Countries (RIS), on June 15, 2011.
42. Prabhat Upadhyaya, "Is Emission Trading a Possible Policy Option for India," *Climate Policy* (New Delhi: Earthscan, October, 2010), p. 567-568.
43. Email correspondence with Dr. Matthew Paterson, University of Ottawa, on August 06, 2011.
44. Vijay P. Ojha, "International Trading of Emission Rights: Its implication for Low-Carbon Growth in India," *India Infrastructure Report 2010: Infrastructure Development in a Low Carbon Economy* (New York: 3iNEtwork Infrastructure Development Finance Company, Oxford University Press, 2010), p. 145.
45. For more information, see http://gain.globalai.org/ranking/readiness, accessed on October 05, 2011.
46. Swaminathan S. Anklesaria Aiyar, "India, World Leader in Natural Disasters," *The Times of India* (May 14, 2006), see http://timesofindia.indiatimes.com/home/opinion/sa-aiyar/swaminomics/India-world-leader-in-natural-disasters/articleshow/1528671.cms, accessed on October 19, 2011.
47. Ibid.
48. See, "Cost of Natural Disasters US$ 109 billion in 2010: U.N.," *Reuters*

(January 24, 2011), http://www.reuters.com/article/2011/01/24/us-disasters-un-idUSTRE70N26K20110124, accessed on October 19, 2011.

49. Interaction with Dr. Kirit Parikh, Chairman, Expert Group on Low Carbon Strategies for Inclusive Growth, at Integrated Research and Action for Development, New Delhi, on July 19, 2011.
50. Interaction with Dr. Arabinda Mishra, Director, Earth Science and Climate Change Division, at The Energy and Resources Institute (TERI), on July 14, 2011.
51. Email correspondence with Dr. Matthew Paterson, University of Ottawa, August 06, 2011.
52. Interaction with Celo Paskal, Associate Fellow, Chatham House, at Manipal University, on September 20, 2011.

6

The Politics of Environmentalism: Relations Between State and Non-State Actors

> "We have wished, we ecofreaks, for a disaster or for a social change to come and bomb us into Stone Age, where we might live like Indians in our valley, with our localism, our appropriate technology, our gardens, our homemade religion -- guilt-free at last!"
>
> — Stewart Brand, Editor, Whole Earth Catalogue

Environmentalism is a broad social and philosophical movement that has redefined human history on account of not only its enormous reach but also its future repercussions. Today, protection of environment has become the top-most 'priority' of every actor in the world, both state and non-state, due to internal and external pressures that may not culminate in the 'common good'. The importance of the concept has enhanced since the identification of global warming and climate change as the biggest threats to humankind. India is one of the most vulnerable countries, and hence, has to find solutions to tackle the effects of environmental change as well as take steps to stop it to the greatest extent possible. This endeavour would indeed require the State to join hands with non-state actors to not only devise solutions but also implement them effectively. Among non-state actors, the Environmental Non-Governmental Organisations (ENGOs) have a significant say as far as environmental governance is concerned. They play a prominent role in formulating a country's stand on environmental issues at the international level, plus, they act as a bridge between the Government and the public. Instead of acting as a bridge, these organisations could also be at loggerheads with the Government justifiably or unjustifiably if the latter fails to take the former and the people into confidence both with good governance as well as astute public diplomacy efforts. Just as the nation states indulge in the politics of environment, the ENGOs also do the same. The advent of globalisation (or the emergence of a global world) and information communication technologies has also given impetus to the environmental movement besides posing new challenges to the State and non-state actors. In this context, this chapter strives to trace the origin and evolution of environmentalism. The strategies adopted by the environmental groups are as important as the phenomenon of environmentalism itself. They have both been a source and product of environmentalism. The question of whether or not modern environmentalism has led to the emergence of a global civic society and a simultaneous roll back of the State is also pertinent. This is addressed in detail in the chapter by looking at environment from the perspective of the State, the civil society as well as the entire globe.

The chapter discusses the impact of environmental groups on policy formulation and implementation at both the international as well as national levels. In this respect, it analyses the challenges of governance and accountability to environmental movements by highlighting specific examples of ENGOs. They have played both positive and negative roles, the former by raising awareness as well as assisting the establishment in data generation and dissemination, and the latter by misinforming the public and the Governments as well as indulging in corrupt practices for vested interests. A comparison between the West's treatment of ENGOs and NGOs dealing with environmental matters and that of India would give an interesting background to the lack of strategic thinking in India's policy makers. In this context, issues such as economics, strategy and diplomacy take the centre stage. The Corporate-ENGO confluence and lack of transparency on the part of the Government of India are some of the factors that have provided an impetus to the increasing role of the ENGOs in environmental governance in the country. Their influence on the Government could be both progressive and detrimental. These aspects would be closely studied in this chapter encompassing both governance and strategy. The attempt has been to deal with every issue mostly from the State's perspective although other perspectives are provided adequate focus simply because the State does occupy the core of international relations to this date. Also, the chapter does not touch upon any specific policies of any specific Government. The focus is more on a classical liberal democratic setup and the recommendations are also directed towards the State and in the process, also to the ENGOs.

Origin and evolution of environmentalism

Different Facets and Definitions of Environmentalism

To define environmentalism as environmental conservation would be too simplistic, as it has different layers such as social, political, economic, scientific and even philosophical. If one explores the evolution of environmentalism and its current status, one could say that the State-ENGO relations form a critical part of international environmental affairs. The different facets of environmentalism have been also moulded by the emergence of economics as a very important domain in the international arena. The social aspect of environmentalism is largely guided by activists, lobbyists and environmental groups that try to influence the political process as well as individual behaviour. By this, they try to promote policies and actions at the State and the individual levels respectively emphasising on the importance of sustainable management of natural resources along with protection of ecosystems. A good example would be the environmental groups that strive to educate the masses about the need for conservation of water or electricity by working in the grassroots. At the same time, States are increasingly seeking recommendations of these environmental groups on issues such as environmental clearances or framing of climate change policy. The political nature of environmentalism has surfaced in various corners of the world as environmental groups have managed to challenge the State's credibility over the management of environmental issues and overturn some of the State's policies. It is claimed by certain environmentalists that these groups have given the victims of environmental disasters a separate political identity by coming to their rescue while the State failed to fulfil its obligations towards its citizens. A classic example of this is the Bhopal gas tragedy of 1984 that took millions of lives and continues to affect the population in the adjoining areas; the State failed to live up to the expectations of the local population that drove the latter towards the emerging humanitarian groups that became political actors, and gave them an alternative. Such initiatives could be imbibed by environmental groups as well. Today these groups have assumed the role of their representative against the State. The economic strand of environmentalism is crucial as the success of any of these movements is heavily dependent not only on human resources but also on sufficient economic resources for their sustainability. The sources, management and utilisation of these resources are keys to the legitimacy of environmentalism. Very often environmentalism is considered to be divergent to economics as economic activities are written off as anti-environmental; this assumption derives its crux from the direct linkage that exists between economic development and environmental degradation (consumption of resources and pollution). The scientific facet of environmentalism is the obvious equation of environment with its physical, chemical and biological features, thereby stressing on the alterations in those features which in turn create the platform for environmentalism to flourish. Last but not the least, the philosophy of environmentalism is as old as the inception of the Vedas and the Bible. As per the Rig-Veda:

> *"O mother earth let thy bosom be free from sickness and decay*
> *May we through long life*
> *Be active and vigilant*
> *And serve thee with*
> *Devotion"*

Some thinkers such as L. Buell conceptualises environmentalism as 'an umbrella term that may stretch to cover any environmental reform movement, whether anthropogenic or ecocentric, radical or moderate, although some activists, such as environmental justice advocates, would want to dissociate themselves from any implication that environmental ills are more fundamental than social ills such as poverty and racism'. T. Benton goes one step ahead and deems environmentalism as political philosophy that covers 'a range of social and political movements and value perspectives which share a concern to protect or improve the quality of the rural, urban, domestic or working contexts of modern social life'.[1] Whichever definition is taken into consideration, the fact is that no issue is single-faceted. One more factor that has to be acknowledged is that environmentalism is often considered a school of thought that is human-centric. It is ecologism that is environment-centric; it is based on the belief that nature should not be exploited whether it is of any value to humans or not. Therefore, when one talks about environmentalism, one has to keep in mind that only those aspects of environment that affect human life or survival are encompassed in it.

Environmentalism in the West

Environmentalism is not a new phenomenon and such a fallacy must be discarded in order to understand it better. It has been an integral part of human history since time immemorial in different forms. Take the example of the Waorani tribe that still lives in seclusion in the Amazon rainforest for thousands of years, and has believed in sustainable living. The Bishnoi tribe of Rajasthan, India has made ecological sensibility and protection of flora and fauna cornerstones of their religion. Even in the 17th and 18th century, the French and the Dutch colonists imposed forest conservation and water pollution control policies in Mauritius. The British poet, William Wordsworth, inspired an entire generation and continues to inspire young minds with his pantheistic philosophy. His poems such as 'Daffodils', 'Calm is all Nature as a Resting Wheel', 'Animal Tranquility and Decay by the Sea', 'The Nightingale', 'The Skylark', 'Written in Early Spring', 'Yew-trees', and others celebrate nature to the extent that it is regarded as a human with its own body, soul and mind. Other instances include writings concerned with environmental pollution by writers such as Alkindus, Costa ben Luca, Rhazes, Ibn Al-Jazzar and others in the Arab medical treatises during the Arab Agricultural Revolution. The birth of modern environmental movement could be traced back to the industrial revolution in Europe characterised by the emergence of gigantic factories besides uncontrolled consumption of fossil fuels. During this period, various environmental movements sprung up that forced the establishment to enact environmental laws such as the British Alkali Acts (1863) that regulated gaseous hydrochloric acid emitted during the production of soda ash (Leblanc process). Thereafter, plenty of conservation-based groups such as the Commons Preservation Society, the Kyrle Society, the Royal Society for the Protection of Birds (RSPB) and the Garden city movement emerged in the UK. Similarly, national parks were created in Australia, New Zealand and Canada.

In the US, environmentalism took shape in the 1850s with the publication of Henry David Thoreau's work, 'Walden' (1848) and John Muir, a naturalist who founded Sierra Club in 1892. Both influenced the Government with their literary and philosophical works. Benjamin Franklin's successful attempt of removing tanneries from Philadelphia's commercial district and founding of the National Park Service by Woodrow Wilson in 1916 were some of the other highlights of environmentalism in the early 19th century. Endangered wildlife also became focus of attention with the extinction of species such as passenger pigeon and the North American buffalo. These examples clearly show that environmentalism was considered synonymous to conservation until the 1950s. There was a marked shift from conservationist principles to larger issues related to environment when 'Silent Spring', a work on the environmental impacts of indiscriminate spraying of DDT in the US, by Rachel Carson, an American biologist, was published in 1962. The book managed to galvanise public opinion against the use of DDT that could have disastrous implications for both human health and ecology. This led not only to the creation of the US Environmental Protection Agency in 1970 but also to the banning of the agricultural use of DDT in the US in 1972. This was a watershed development that had taken environmentalism to a new level and the role of the civil society in policy formulation was bolstered. Thus, the real movement is said to have begun only in the 1960s. Also, this gave impetus to the rise of more pressure groups such as Greenpeace and Friends of the Earth (FoE) in 1971. These groups again brought to the fore the importance of protection of threatened species like pandas and tigers and also tried to bring to an end trade of elephant ivory, rhino horn and seal fur. The mandate of these groups has since then expanded to multiple facets of environmental protection including energy security. The 1970s witnessed the beginning of international/inter-governmental cooperation, which would be touched upon while discussing modern environmentalism.

Types of Environmentalism

Radical Environmentalism

During the last three decades of the 20th century, environmentalism underwent further change with reemphasis being laid on environmental ethics, giving shape to a more radical form of environmentalism, referred to as 'deep green ideology', which is different from mainstream environmentalism. This view was propagated by environmentalists who had prophesied an environmental catastrophe, understood life on earth as a single organism as well as merged their movements with political issues such as the anti-Vietnam war protests. This ideology never got inducted into mainstream environmentalism that laid emphasis on enforcement of more laws and regulations for the protection of environment. Bron Taylor describes radical environmentalism as a movement that 'most commonly brings to mind the actions of those who break laws in sometimes dramatic displays of 'direct action' in defense of nature. Such action—which may involve civil disobedience and sabotage, some instances of which involve risks to people —have generated criticism and vilification of these movements. Radical environmentalists are sometimes labeled terrorists and believed to harbour, and hope to develop, weapons of mass death'. Radical environmentalism is widely practiced by groups like Earth First! and The Earth Liberation Front, apart from bioregionalists and green anarchists, deep ecologists and ecopsychologists, ecofeminists and participants in the feminist spirituality movement, Pagans and Wiccans, anti-globalisation protestors and some animal-liberation activists.[2]

Modern environmentalism

According to a particular group of scholars, the growth of modern environmentalism is said to have gained impetus on account of four major events—first, the publication of the 'Limits to Growth' by the Club of Rome in 1972, which famously declared that the only way to decelerate the growth in the natural resource depletion, pollution and population would be to halt growth itself rather than using technical innovations to combat the negative upshots of growth; second, the UN Conference on Human Environment held in Stockholm in 1972, which got wide publicity as a result of participation of 113 nations, 19 inter-governmental committees and about 400 non-governmental organisations; third, the World Commission on Environment and Development established by the UN in September 1983 that engendered the all-important debate on sustainable development through a report submitted by an independent commission headed by G. H. Bruntdland in 1987, and fourth, 'The Earth Summit', held in Rio de Janeiro in 1992 that was the largest conference so far held in world history with 178 government delegations, 117 Heads of States and 1400 non-governmental organisations.[3] They also highlighted the role played by the environmental pressure groups like the FoE, Sierra Club, World Wildlife Fund (WWF), Greenpeace and others besides the 'Green' political parties in the growth of modern environmentalism in most of the countries of the world.

New Environmentalism

It was in the 1990s that environmentalism turned into an enormous force that began to not only create awareness using information communication technologies and social media but also make their presence felt across the world cutting across territorial jurisdictions. This was boosted further by emerging threats of global warming and climate change that hold the potential to devastate not just one country or a group of countries, but the entire globe. In the 21st century, environment has become the buzz word and the emergence of 'new environmentalism' is a sign of that. There are various trends that help in understanding 'new environmentalism' better —first, from a movement that was largely dominated by idealistic, elitist, urban, young men, environmentalism has transformed into one that cuts across generations, gender, region, partisan lines and income groups; second, changing weather patterns, that is climate change on account of global warming, has become the rallying point for the new environmentalists; third, global warming is increasingly being perceived as a symptom rather than a problem; the problem in fact lies in the wasteful nature of the society, and fourth, these changes in attitudes are being driven by the rising energy prices, which in turn has steered the consumers to reduce expenditure on energy in a great way.[4] Energy security has become an issue of concern for all the countries, especially the developing ones.

Evolution of Strategies of Environmental Groups

Another important factor that has brought the ENGOs to the forefront of environmental policy-making is their institutionalisation. This has been most profound in Northern America and Western Europe, especially in the Nordic countries, Germany and the Netherlands. Conservation groups mostly resort to lobbying, dissemination of information and taking refuge in the legal system to protect the environment. These groups have become institutionalised in terms of membership, income, professionalisation as well as access to the policy process. For example, Sierra Club campaigns include 'global warming programme', 'smart energy solutions' and safe and healthy communities'. These groups have acquired significant influence within the policy process of the Western countries by engaging in regular dialogue with the officials and bureaucrats. As far as international organisations are concerned, institutionalisation has been comparatively slower due to their close association with radical social movement. Both Greenpeace and FoE initially believed in using direct action, that is confrontational methods, usually within the law. Greenpeace shot to international fame in 1985 during the Rainbow Warrior incident in which its ship was blown up by the French Government agents and one of its crew members was killed. The ship was used to protest against the French nuclear testing. Off late, FoE has stopped using direct action as a strategy due to the threat posed by the courts against its financial assets, whereas Greenpeace is still committed to the use of direct action. However, it has also added new strategies under its belt such as 'solutions-led' approach by conducting research on key issues, disseminating information and appointing scientists or people with scientific temper to crucial posts.[5] It is this approach that has driven Greenpeace into forging partnerships with corporations.

Modern Strategies of Environmental Groups

The different methods used by ENGOs to influence policymakers include demonstration of the sheer size of support, causing public disruption and economic damage to business or Government and the famous 'bearing witness' strategy employed by Greenpeace on various occasions. However, in the age of information technology revolution, Internet has emerged as an important medium of bringing about social change. Environmental organisations and activists have also capitalised on the increasing influence of cyber space on the civil society. Environmental movements have been a direct outcome of lack of accountability and efficiency on the part of the political syndicate in addressing environmental issues. Another reason for the magnifying stature of the ENGOs is lack of transparency on the part of the State. Take the case of even a country like China that releases white papers quite frequently in the public domain; this culture is absent in the Indian system as of now. Environmental movements were mostly spearheaded by a few individuals, and it is their imperative to collect and propagate information for two reasons – first, any movement dies if it does not have sufficient support base that could be galvanised through diffusion of 'credible' information only, and second, issues such as that of environment are perpetual and thus, it is essential for the groups to build a firm and reliable information foundation for the future generations to follow. Apart from traditional tools such as telephone, mass mailings and rallies, web tools are being widely used by these groups to propagate information and create awareness. 'Online strategies' have made the task of these groups easier as they help them overcome lack of resources as well as that of recognition in the political sphere and in other spheres. It is the grassroots environmental groups that utilise them more than the mainstream ones as they generally tend to get marginalised in the society. As Laurie A. Kutner points out, "For activist organisations with limited resources, Internet-based technologies are providing fast, easy, and cost-effective means through which to access, use, create, and disseminate information."[6] They also give vast gamut of opportunities to citizens along with non-citizens cutting across all sections of the society, especially in terms of granting certain amount of power to the powerless entities. It has broken down social, physical, geographical and political barriers on account of various reasons. Grassroots environmental organisations interact with regional, national and international organisations. Furthermore, groups working for different causes do collaborate their actions in order to reach out to a wider audience, and to find more holistic solutions to environmental issues. They have also started to use new media as a vehicle by which they could directly criticise as well as give recommendations to the political authorities. The same groups that were earlier constrained by their inability to access the mainstream mass media are now feeling empowered by the advent of internet-based technologies especially online networking, known as social networking sites in other words. One characteristic of the internet that has marginally intensified participation of not only the environmental groups but the civil society in general in the political process/ policy-making process is disintermediation. Not only does it facilitate uninterrupted, unmediated flow of political information thereby changing the patterns of political participation, it also 'allows for many messages to be delivered to many small audiences (a concept termed 'narrowcasting') rather than a few messages delivered to few large audiences (broadcasting)' as asserted by M. S. Bonchek. Interestingly, it has also been the target of the State actors for its use by the NGOs to sway public opinion in their favour though counter-information. This leads to conflict between the State and the ENGOs; Pentagon has in fact dubbed this phenomenon 'netwars'.[7]

Ecological Citizenship: All for One, One for All

The Prospects for the Emergence of a 'Global Civic Society'

The understanding of ecological citizenship is crucial for the understanding of the formation of a truly global civic society. One of the causes of the emergence of global citizenship has been the global nature of the problems that the collective humankind faces. Some would suggest that instead of replacing traditional citizenship with global citizenship, it would be more appropriate to transform traditional citizenship to a 'new multi-layer citizenship' as States continue to be the supreme actor in the international arena. Bart van Steenbergen suggests two types of global environmental citizens: "The earth citizen, who is aware of the earth as a living organism and her origin from earth, which leads to the concept of care instead of subjugators – and the world citizen who perceives of the environment as a matter of 'big science' and the planet an object of global management which requires large-scale organisation and government."[8] Ecological citizenship goes one step ahead to recognise non-human beings as 'fellow citizens'. Since they cannot be incorporated in the political decision-making process for practical reasons, ecological citizenship requires human beings to take the responsibility of protecting them. Environment is one of the factors that unite the entire world and its population and therefore, the concept of global environmental citizenship invites plenty of attention. Marshall McLuhan's concept of 'global village' talks about integration and evolution of global values that could make the world a better place to live in with much more complex interdependence along with risk-sharing partnerships. The term citizenship connotes the existence of the State. Global citizenship debunks national identities and national boundaries. However, it has been seen that very few actors – State or non-State – have shown the willingness to defy national sovereignty in favour of environmental issues.

Ecological citizenship has three tenets—ecological awareness, ecological citizenship as a legal concept, and ecological citizenship as a political concept. Ecological awareness implies that a citizen must 'understand the importance of making collective decisions regarding the commons, seek the common good, identify with bioregions and ecosystems rather than obsolete nation-states or transnational corporations, consider the wider impact of his/her actions, be committed to community building, observe the flow of power and act according to his/her convictions'. The other two aspects are described in detail as rights in Principle 13 of the Earth Charter (an international declaration of fundamental values and principles considered useful by its supporters for building a just, sustainable, and peaceful global society in the 21st century). These rights aim at 'strengthening of democratic institutions at all levels', and providing transparency and accountability in governance, inclusive participation in decision-making, and access to justice'.[9]

These rights very clearly reinforce the growing relevance of public diplomacy in the twenty-first century. They call for direct public participation in global affairs, strengthening of domestic institutions, democratisation of information, strengthening of local communities and ensuring freedom of individuals. States can no longer act as black boxes and at every level they have to compete for credibility with different parties such as media, corporations, NGOs, inter-governmental organisations, networks as well as scientific organisations. On dissecting them further, one could clearly see that the first tenet ('Uphold the right of everyone to receive clear and timely information on environmental matters and all development plans and activities which are likely to affect them or in which they have an interest') stresses on sharing of information with the public. The second and third tenets ('Support local, regional and global civil society, and promote the meaningful participation of all interested individuals and organisations in decision-making', and 'Protect the rights to freedom of opinion, expression, peaceful assembly, association, and dissent') strive to empower the civil society by championing its participation in the formal process. The fourth and the fifth tenets ('Institute effective and efficient access to administrative and independent judicial procedures, including remedies and redress for environmental harm and the threat of such harm', and 'Eliminate corruption in all public and private institutions') emphasise on the importance of accountability and transparency at all levels. The sixth tenet ('Strengthen local communities, enabling them to care for their environments, and assign environmental responsibilities to the levels of government where they can be carried out most effectively') openly calls for diffusion of governance so that the tasks could be carried out more effectively.[10] However, these rights are balanced by duties just as the fundamental rights in the Indian Constitution are balanced by the fundamental duties.

The State's Response

While talking about environmental degradation, it is equally important to analyse environment from the perspective of the State. Environment could be divided into various categories based on ownership, exploitation and responsibility; and the State has to act according to how the environment is defined by it. There is one environment which is owned by the State whose jurisdiction is restricted to just that particular State and no other State would have any rights over it. This is applicable particularly to the natural resources available within the boundaries of that nation state. In short, this environment could be used, overused, exploited, replenished, protected, preserved and destroyed according to the whims and fancies of the State that owns it. The second category of environment is usually the one that is owned by other States. The State does not have sovereignty over this environment and therefore, it cannot use it to fulfil its interests ideally.

Unlike the first category of environment, the second one cannot be exploited or destroyed. If it does, it is expected to create an atmosphere of tension and conflict between the two States. On the one hand, it has an obligation not to cause any damage to the second environment; and on the other hand if it does, the repercussions could be felt even within the State especially if the other State is the former's neighbour. If the resources are shared between the two States such as in the case of sharing of river waters, conflicts could arise if political agreements are not reached upon and not followed in true spirit and letter. The third category of environment is owned by all. What differentiates this environment from the previous one is that in the idealistic sense, no State has the right to cause any harm to this environment, but unlike the third category of environment, the political repercussions of meddling with this environment are negligible or even nonexistent. The fourth category of environment is owned by none. The line between the third and the fourth is very thin. One could claim that the atmosphere is collectively owned by all nation states but is actually owned by none. If one accounts the various pollutants being released into the atmosphere, then the question of responsibility arises. The efforts are, thus, directed towards fixing the responsibility and make the atmosphere a collectively owned entity as ownership leads to responsibility. In this context, anything that affects the atmosphere is also said to have a direct bearing on the climate.

The Greenhouse Gas (GHG) emissions released into the atmosphere are considered to be the cause for the greenhouse gas effect and global warming and in turn, climate change. Therefore, it is very clear that climate change does not respect political boundaries and it affects everyone. The principal factor that distinguishes the third category from the fourth category in a broad sense is that of the 'right to exploit'. For instance, if the atmosphere is declared as an entity that is owned by none, no State would have the right to emit GHGs into it in the idealistic sense. Another interpretation of this category could be that this environment is a free-for-all entity. In such a case, 'responsibility' assumes significance in which case no State would be effectively held accountable for any damage caused to the fourth category of environment as universal jurisdiction is yet to take shape at the international level. Hence, the countries would be allowed to emit as much as possible without any legal or political restrictions turning the atmosphere into an entity that is owned by none and therefore meant to be exploited. In another example, one could include high seas which are outside of national jurisdiction under the third category as they are collectively being used by everyone—under various international arrangements such as the UN Convention on the Law of the Sea (UNCLOS). The chances of the fourth category of environment being claimed for 'single ownership' are more than the third category as it is not owned by anyone. Claims and counter-claims over the high seas are extremely plausible in future with changing environment and climate as well as increasing need for resources. In the case of natural resources that have so far been not discovered could be unearthed in future which would then be a source of conflict between States for ownership. Hence, international cooperation is inevitable and extremely essential for international peace and stability. Even though climate change falls under all the four categories, it has been classified as the fourth category of environment with States not showing commitment towards taking the responsibility for the problem and addressing it collectively.

The phenomena of global warming and climate change also challenge the concept of sovereignty. While state sovereignty continues to be the most decisive factor in international politics, environment debunks state sovereignty to a certain extent. Environment cannot be divided into territories. At the same time, every State has the right to exploit its territorial natural resources as well as a moral (or even to some extent legal) obligation to protect the environment that belongs to the whole world, that which is not restricted to a particular territorial space. Therefore, the States' sovereignty 'allows them to use, but not to abuse, pollute, exploit or otherwise overuse the environment'.[11] Environmental governance and territorial sovereignty are interlinked. This is the reason why to this date the developed and developing countries have not been able to reach a consensus on the issue of 'climate justice'. On the one hand, developed countries insist that every country including the ones belonging to the developing world should take up the burden of fixed emissions cuts. On the other hand, developing countries continue to raise the issue of their rights to achieving development which developed countries have already achieved in the process of which they contributed most to GHG emissions. As Klauss Bosselmann comments, "Ecological governance conceptually restricts territorial sovereignty leading to a paradigm shift in international environmental law: not state sovereignty sets the limits to environmental protection, but environmental protection sets limit to state sovereignty."[12]

While looking at environmental issues from a global perspective, the role of international institutions in global environmental governance is eminent. It is a fact that inter-governmental organisations such as the United Nations Framework Convention on Climate Change (UNFCCC), the Inter-governmental Panel on Climate Change (IPCC) and others have been formed under the aegis of the UN; there are regional organisations that have also been making efforts in the areas of adaptation to and mitigation of climate change. Yet, these organisations have proven to be more or less incompetent in dealing with the multifarious, interdependent and political nature of environmental problems. This will be dealt with in the next chapter. The creation of networks of governance has definitely added another layer to global environmental governance, but still the nation states have monopoly over these international institutions. Interests of the States overrule the needs of global environmental governance. Even the Rio Declaration of 1992 plainly defends state sovereignty over natural resources in the Preamble. Neither have certain States relented to the core tenets of the Kyoto Protocol for different reasons. Besides, UN Environmental Programme (UNEP) set up by a resolution of the General Assembly does not have any executive powers; its objectives are completely dependent on persuasive powers. The current governance on global environmental issues is flawed as an enforcement body is absent.

The Interactions and Counter-interactions in a World of Multiple Relationships

The Prospects for Environmentalism and Diffusion of Governance in a Liberal Democracy

As one talks about democratisation of institutions by increasing public participation, one also needs to closely analyse the persisting view that maintains that sustainability could be achieved though democratisation. This implies that somewhere down the line, democracy facilitates and probably enhances protection of environment. In fact, principle 10 of the Rio Declaration states that 'environmental issues are best handled with the participation of all concerned citizens, at all relevant level'.[13]

Environmental politics is one of those rare domains in which the policy-making has indeed witnessed wider participation from the civil society, especially the ENGOs. The role of the ENGOs assumes greater significance in a liberal democracy as they 'identify risks, assess environmental impacts and design and implement measures to deal with them, and maintain the high degree of public and political interest required as a basis for action'.[14] However, the ENGOs have also faced criticism from various quarters on two grounds – first, the ability of the ENGOs in influencing the promotion of environmental sustainability might be overstated though they provide for popular participation, and second, the democratic nature of the ENGOs has been viewed by many with scepticism primarily because of the fact that with increasing institutionalisation of these ENGOs, the participatory tradition is giving way for a more elitist or hierarchical tradition. The activities of the ENGOs are multifaceted—first, they might take on the role of agents of policy-makers by carrying out the task of implementing the policy-makers' agenda in the society. This activity is also aided by the Government through funds; second, the ENGOs could take the views of the citizens of the society to the policy-makers, thus representing either their supporter base or the general public, and third, the ENGOs also play an important role in terms of transforming the behaviour and attitudes of the citizens towards environment. In this regard, besides engaging themselves in creating publicity for their cause and galvanising communities or even joining hands with the existing communities in order to manage local environment, the ENGOs have also begun to promote non-governmental environmental governance by mobilising support for their visible objectives and indirectly influencing the government policy. As James Rosenau said, "It is possible to conceive of governance without government—of regulatory mechanisms in a sphere of activity which function effectively even though they are not endowed with formal authority."[15] Fourth, as the ENGOs take the citizens' views to the policy-makers, they also attempt to change the views of the policy-makers through various means such as lobbying, participation in formal environmental institutions, public campaigning, and so on.

It is also highly important to analyse the strategies of the ENGOs to disseminate an 'ecological sensibility' among the masses in a liberal democracy. Some organisations produce educational material and indulge in civil disobedience or campaigning primarily to attract media attention. Others such as Greenpeace and FoE, invest considerable amount of their resources in scientific research in order to directly present scientific evidence to the citizens as well as the Government. At the same time, as discussed earlier, organisations such as Greenpeace have gone to the extent of indulging in actions that 'excite the eye'. Paul Wapner comments, "By offering spectacular images to the media, Greenpeace invites the public to bear witness; it enables people throughout the world to know about environmental dangers and tries to pique their sense of outrage."[16] The ENGOs come in direct confrontation with the liberal democracy when they engage in violent stunts to make their voices heard or to seek attention as it would be deemed violation of justice. Similarly, if the ENGOs protest against the moral wrongs of the Government and not injustices, the ENGOs tend to lose their credibility in a liberal democracy.

Government-ENGO Partnership

If one delves more deeply into the role of the ENGOs as independent agents undertaking various steps to influence the government policy, one could clearly see that the ENGOs contribute not only their views but also the citizens' views to the policy process. Representation is an important factor that forces Governments to take them with sufficient seriousness. Apart from representation, they are also approached by the Government for their expertise as they could provide the latter with relevant information which the Governments would otherwise would have never come to know. As already mentioned earlier, the gap between the ENGOs and the scientific community is slowly reducing as the ENGOs themselves are devoting considerable amount of their resources to social science and natural science research in environmental issues. For example, The Energy and Resources Institute (TERI) have been focussing on research activities related to climate change, efficient utilisation of energy, sustainable development and large-scale adoption of renewable energy technologies. It has also played a pivotal role in formulating the developing world perspective in the international climate debate apart from providing vital inputs for the IPCC process. The institute's areas of focus in climate change research are 'impacts and vulnerability assessment, adaptation strategies, exploring GHG mitigation options and issues therein, climate change policies and climate modelling activities'.[17] Internationally, it is involved in several collaborative projects with institutions in other parts of Asia, the Asia Pacific, Europe and Americas. In fact, it is already on the path of making efforts to help Africa build capacity to mitigate and adapt to climate change. The majority of reports and documents of the Ministry of Environment and Forests (MoEF) and the Planning Commission are replete with data generated by TERI, whether it is the sector-wise inventory of GHG emissions or even policy recommendations in terms of options and potential for GHG emissions mitigation. The Interim Report of the Expert Group on Low Carbon Strategies for Inclusive Growth of the Planning Commission is one of them.[18] Another example of an ENGO in India that works closely with the MoEF is Development Alternatives (DA). It was chosen as the Environmental Information System (ENVIS) node # 4 by the MoEF. It is linked to the global environmental information system called INFOTERRA/UNEP. ENVIS # 4 has been 'assigned the task of collecting, organising and dissemination of information, globally on Environmentally Sound and Appropriate Technologies (ESAT)'. Its primary objective is to create databases and update them.[19] It not only releases timely information to the public free of cost but also assists the Government in generating data which could be used both in the domestic as well as international setups such as for the Nationally Appropriate Mitigation Action (NAMA) documents that are submitted to the UNFCCC. The entity has always worked towards rural habitat policies and promotion of green jobs in a great manner. These organisations, therefore, have been influencing the energy policies of the country particularly in terms of diversification of the energy mix and enhancing the share of alternative sources of energy.

Impact of the Environmental Non-Governmental Organisations at the International Environmental Negotiations

The role of the ENGOs has expanded to the arena of international environmental institutions. Earlier, the NGOs were given limited access to international negotiations on environment and they were seldom acknowledged in the multilateral environmental treaties. The criteria for inclusion of the NGOs were very rigid as the ENGOs were expected to be qualified in a particular field such as protection of ozone layer or fields related to hazardous wastes. The emergence of a revitalised global civil society is largely being attributed to the robust NGO activity. Therefore, there is an existing view that the General Assembly of the UN must be revamped in order to include the corporations and people and their associations besides the Governments. The States have had an upper hand in State-ENGO relations despite the fact that the NGOs have also been successful in enhancing their visibility and power through their specialised and useful resources. States have eclipsed the role of the ENGOs at all the negotiations by being highly 'realistic' in their approach in comparison to the highly 'idealistic' approach of the latter. Since the State is the highest decision-making authority in the world, its position on all issues related to environment has prevailed over that of the ENGOs. Among the prominent NGO constituencies in the UNFCCC process include BINGO (Business and Industry NGOs), ENGO (Environmental NGOs), TUNGO (Trade Union NGOs), IPO (Indigenous People's Organisations), LGMA (Local Government and Municipal Authorities), RINGO (Research-oriented and Independent Organisations), YUNGO (Youth NGOs), Faith (Faith-based NGOs), and Gender (Gender-based NGOs).[20]

The role of the ENGOs at the international environmental negotiations took shape at the first Climate Change summit, commonly referred to as the Rio Summit in 1992. They deserve to be acknowledged for their efforts to bring to the attention of the world community, the impending disaster that was awaiting them and to compel the world leaders to devise concrete measures in this regard. In her article, 'Rio Diary: A Fortnight at the Earth Summit', Fiona Godlee reports, "Their (campaigners from Greenpeace and three other NGOs) 10-point plan to save the summit from failure included the things they felt had been left off the agenda: legally binding targets and time-tables for reducing greenhouse gas emissions, a reduction in consumption by the North, global economic reform to improve the terms of trade for the South, more financial commitment from the North and an end to the World Bank's control of the Global Environment Fund, a ban on the export of hazardous waste, recognition of the rights of indigenous peoples, and an end to nuclear weapons and nuclear power."[21] At the end of the Earth Summit, the parties came up with a treaty that was legally non-binding that set no mandatory limits and contained no enforcement mechanisms, quite contrary to the pleas of the ENGOs.

In order to gauge the degree of influence of the ENGOs on international environmental organisations, Elisabeth Corell and Michele M. Betsill outlined seven parameters in one of their essays. They are "(1) being present at the negotiations; (2) providing written information supporting a particular position (such as newsletters, research reports or papers, or information leaflets) to relevant government ministries or to the negotiation sessions; (3) providing verbal information supporting a particular position (through statements, information meetings or seminars during negotiation sessions); (4) providing specific advice to government delegations through direct interaction; (5) opportunity to define the environmental issue under negotiation; (6) opportunity to shape the negotiating agenda, and (7) ability to ensure that certain text supporting a particular position is incorporated in the Convention."[22]

The role of the ENGOs in the signing of the Kyoto Protocol has been investigated by many analysts. The influence of the ENGOs in this case was highly contingent on the intangible nature of the issue, the delineation of highly technical solutions, the politically and scientifically driven history of the issue, the framing of the issue based on economics and so on. The political opportunity for the ENGOs at the Kyoto Summit was highly restrictive. Interestingly, the summit was dominated by the ENGOs from the North (developed world) despite the presence of a significant number of them from the South (developing and under developed world). The four major objectives of the ENGOs during the Kyoto Protocol negotiations were that the Protocol should (1) require industrialised countries to reduce their GHG emissions 20 per cent below 1990 levels by 2005; (2) include strong review and compliance mechanisms; (3) not allow industrialised parties to meet their commitments through emissions trading, and (4) not allow parties to get credit for emissions absorbed by sinks. None of these recommendations were included in the Protocol. The commitments made by industrialised nations were least stringent especially since state delegations doubted the feasibility of a 20 per cent reduction target. There was no scope whatsoever for any compliance or review mechanisms. Article 17 and Article 3 of the Kyoto Protocol quashed the last two objectives of the ENGOs by permitting the States to do emissions trading and to get credit for emissions absorbed by sinks respectively. From this example, it is evident that though the ENGOs undertook numerous activities throughout the Summit under the umbrella of the Climate Action Network, they could hardly make any impact on the final outcome. It is said that though the ENGOs did not have direct access to the plenary or closed-door sessions, their representatives were in constant touch with selected delegates who sought their opinion on various issues, thus giving them the opportunity to participate in the debates indirectly. In terms of resources, they had superior technical knowledge compared to the other participants including Government delegates. Another privilege that the ENGOs enjoyed during the negotiations was the backing of the public. The biggest achievement of the ENGOs was their success in shaping the debate over emissions trading by coining the term 'hot air' (referring to the ability of a country whose emissions are already below its legally binding limit to trade the difference). The insertion of sinks into the Kyoto Protocol was also debated extensively due to apprehensions raised by the ENGOs.[23]

Another example of the role of the ENGOs and the civil society at the international environmental organisations is the Klimaforum or People's Climate Summit held at both Copenhagen and Cancun in 2009 and 2010 respectively. At both the summits, nearly 50,000 members of the civil society called for 'System Change, and not Climate Change'. At Klimaforum09 held on the sidelines of the Copenhagen Summit, nearly 202 debates, 70 exhibitions, 43 films, 16 concerts and 11 plays from all over the world were organised. Klimaforum10 was again a grassroots initiative from Mexico during the UNFCCC conference in Cancun, which gave a fruitful platform for the people, grassroots movements and the NGOs to voice their opinion over climate change issues and add to the seriousness of the discussion at the official forum. Both the summits came up with 'a

People's Declaration from Kilmaforum09' and 'a People's Declaration from Klimaforum10' that were submitted to the Conference of Parties. However, the States deemed their demands too far-fetched at that stage and therefore, disregarded most of them. Some of them included "phasing out fossil fuel, reparations and compensation for climate debt and crimes, an immediate global ban on deforestation of primary forests, opposition to purely market-oriented and technology-centred solutions, equitable tax on carbon emissions', etc.[24] Personalities such as Vandana Shiva, founder of Navdanya, Nnimmo Bassey, chair of Friends of the Earth International, and author Naomi Klein were the chief architects of these initiatives. There is discrimination between the different non-state actors as well. At Copenhagen, it is alleged that the business-related nuclear and coal groups as well as the RINGOs (to some extent) dominated the discourse while the ENGOs and other independent organisations were absent from the scene as they were completely sidelined.[25] At the end of the day, business interests involved in the global carbon market matter more than anything else associated with the environmental concerns.

Passing the Buck and Stopping it

Governance and Accountability of the Environmental Non-Governmental Organisations

The governance and accountability of the ENGOs came into the spotlight in the 1990s. This development was a direct consequence of their role in the creation of a global civic society that tries to influence the practices and policies of international institutions. It was also due to the supposed 'roll back of the state' that has divided not only the implementation of policies but also the accountability. The ENGOs can only lead a successful campaign if they remain accountable as their primary task is to ensure accountability and transparency in other social and political sectors. Some critics say that the growing engagement between the ENGOs and major corporations has eroded the legitimacy of the ENGOs. Another section believes that this engagement would usher in more professionalism, good governance and accountability into the ENGOs. IT revolution, in the form of electronic communication and shrinking of distances between the donors and the ENGOs, have also been exerting considerable amount of pressure on the ENGOs to be more accountable. It is also important to note that with increasing environmental consciousness and social pressures, the corporate sector has woken up to the realities of Corporate Responsibility for Environment Protection (CREP), which borrows principles from Corporate Social Responsibility. This in turn drives them into participating more actively in the environmental activities of the ENGOs especially in the form of private funds that largely remain unaccounted for. Besides, some of them have opened their own environment offices to enhance their image as an environmentally responsible entity. As a case in point, the JSW-Times of India Earth Care Awards for excellence in climate change mitigation and adaptation is being seen by some with high amount of scepticism due to the amount of pollution created by the JSW companies. Even corporations are now in the public diplomacy mode as this helps them sell their products. Accountability within the ENGOs has come under the scanner for yet another reason—recruitment of managers and bureaucrats—whose view may not be compatible with the traditionalist view of environmentalism. Environmentalism tends to overtake profit and sovereignty while managers and bureaucrats are products of these constructs respectively. In 2003, Washington Post sent shockwaves by publishing high-profile accounting scandals in one of the biggest ENGOs in the world, The Nature Conservancy. Its finances, management and even scientific practices were found to be fraudulent.[26] This was a major setback for the environmental movement.

Greenpeace was accused of manipulating details and making inaccurate claims in the case of the Brent Spar's drilling rig disposal. The Greenpeace activists occupied the platform to prevent sinking of Brent Spar at the North Feni Ridge on the basis of the allegation that it contained 5,550 tonnes of crude oil; later on these claims were rejected by an investigation commissioned by Shell UK and Greenpeace was forced to apologise.[27] The whole episode tarnished the image of not only Greenpeace but also the entire environmental movement to a great extent. In addition, Greenpeace has very often been accused of being an elitist, hierarchical and undemocratic organisation that hardly gives any organisation rights to local groups and individual supporters. In fact, a former Greenpeace activist, Paul Watson complained, "Greenpeace has turned begging into a major corporate adventure."[28] Also, the ENGOs are clearly seeking the leadership of Information Communication and Technology (ICT) sector to generate IT innovations. Greenpeace believes that the catalysis of an energy revolution, which would help developing countries to activate a swift transition from heavy dependence on non-renewable sources of energy to renewable sources, is possible only with the assistance of the ICT sector, primarily because of their potential to provide large scale solutions required to cut greenhouse gas emissions. Ironically, the ICT sector's carbon footprint is not commendable as it is responsible for approximately 10 per cent of global carbon emissions. One of the Greenpeace studies claims that India is on the way to becoming the second largest carbon emitter after China by 2020 chiefly owing to the expected 12 to 16 per cent annual growth rate in the ICT sector in the coming years.[29] This has raised several questions with regard to the methods adopted by the ENGOs in terms of devising solutions. Thus, while environmentalists bank on the ICT sector in devising strategies and technologies to create low carbon economies, they have also been forced to look into the growing challenges emanating from this sector.

The Art of Engagement

It is very clear that most of the NGOs start gaining prominence due to two reasons – lack of governance or that of transparency – on the part of the establishment. In a highly bureaucratic country such as India, every process is highly time-consuming; this is one of the reasons why the Government very often falls short of showcasing its commitment to the public. This gives the NGOs to fill the void and show the loopholes in the Government's policies besides recommending other options that may or may not be accepted by the Government. A lot of NGOs including the ENGOs have come under the scanner due to ambiguity or anonymity associated with the sources of their funds. It cannot be denied that some of them are run on vested interests. When it comes to funding, one needs to throw light on the possibility of rise of conflict of interests. According to a report by Steven Lawrence, in the US between 2000 and 2008, 'the numbers of grants for combating climate change doubled and the dollar commitment from foundations surveyed increased from less than US$ 100 million to more than US$ 900 million...about one in six U S foundation climate change grants focussed on policy initiatives, with the total giving in 2008 to policy efforts in the range of US$ 112.2 million...Oil and gas industries have contributed US$ 834 million over the past 10 years'.[30] The Department of Defence itself has been backing research associated with climate security, energy security and national security. Even then, reports suggest that funding on the climate security issue is very limited and needs to increase in future.[31] The ENGOs in India are said to be receiving funds from huge multi-national corporations and transnational corporations; very often these foreign entities could even manage to influence the country's decision-making process indirectly through the ENGOs. This could be treated as a breach on the country's sovereignty and such sort of manoeuvrings may not even be in India's national interest. The Government has also been funding projects being run by think tanks and the NGOs. This clearly puts the limelight on the economics of existence which drives these ENGOs towards business proposals from entities other than the Government agencies. Due to conflict of interests, the studies conducted by the NGOs could be biased and they may not be in a position to tell the Government which policy is in the nation's interest and which is not. The Government on its part has been trying to regulate funding by enacting legislations such as the Foreign Contribution Regulation Act (FCRA), 1976 by which any NGO or organisation in India who 'want to receive contribution from foreign source must get permission from the Ministry of Home Affairs'.[32] On the State's part, regulation of the ENGOs or funding them in order to prevent them from releasing publications against the Government may not be considered good strategies in a liberal democracy in the long run. While the other side of the debate would be that these efforts have actually proven to be defunct as the foreign funding, as already discussed, is large.

The word 'commitment' is harped on by individuals and organisations in every sector. It is one thing to 'show' commitment and work towards it; it is a completely different thing to 'showcase' commitment (even while working towards it) and build confidence amongst the larger society. It is a well-established fact that environmental commitment is on top of the agenda of a majority of stakeholders in the contemporary society. Corporate organisations seem to be ahead of the Government agencies in their initiatives to prevent environmental degradation. The CREP has ensured that the same companies, which were shown in a negative light due to their profit oriented practices, indulge in social service that could transform their image. The situation is rather desperate for they understand the adverse impact potential of being seen in a negative light on the environment front when the common man and policy makers alike are concerned of looming global warming. Ironically, every department in and affiliated to the Government has also been doing its bit to advance environment protection, but without a stated policy like the CREP. The only difference between the two is that the former have successfully trumpeted their achievements as an image-building exercise so that their business does not suffer while the latter are yet to publicise their work in the public domain in a comprehensive manner, due to a perceived lack of need. The Central Government has launched the National Action Plan on Climate Change (NAPCC); in the similar vein even most of the State Governments have embarked on this path by launching State Climate Change Action Plans. Since the launch, the Government has not created one forum or website that could showcase the NAPCC's progress. Instead, different agencies and ministries have been highlighting different events or developments on their respective websites; this makes it all the more difficult for the public to keep track of a landmark Action Plan in the history of India. When it comes to the State Action Plans, very few states have taken this project seriously in the first place. Besides, the State Governments' public relations exercises have been restricted to just releasing the official version of the document. In fact, the information displayed on the states' official websites does not cater to the needs of governance. Gujarat is again an exception in this case with fairly well-documented websites, but the coordination between different agencies is still lacking. At a juncture when media attraction is relatively easier to obtain, the Indian establishment should be exploiting these readily available resources to garner support for their initiatives and create goodwill.

As a well-reasoned strategy, the Government departments in the Western countries have used and continue to use their associations with think tanks to hold consultations on pertinent issues related to environment as well as release them for the public to take note of. This custom is absent in India as the association between our handful of think tanks and the establishment is limited to small-scale consultations that are rarely highlighted. These are indications of lack of will to initiate diplomatic missions on the part of the Indian authorities handling environmental matters. The youth in India is largely ignorant of strategic issues that are pertinent to national security of which environmental security is an integral part. There is dire need to involve them through the aegis of these think tanks so that these future leaders have a serious appreciation of the issues related to the country's security when they rise to be the policy makers of tomorrow.[33] For example, in the West defence think tanks (consisting of military personnel mostly) first recognised the need for putting more 'energy' into environmental security related research. This process can only start if the Government accepts the role of defence think tanks and consults them for policy formulations. Centre for Air Power Studies is the only defence think tank in India that has a serving military officer working in this area. In India, the engagement between the MoEF and the various think tanks/NGOs is acknowledgeable as already explained earlier in the chapter. India's international stand on environmental issues is shaped greatly by these organisations. Even in the case of the MoEF, many claim that it is with Jairam Ramesh's induction only that the role of the think tanks/NGOs got augmented.[34] Therefore, there is no reason that the process cannot be emulated by all the other ministries that also deal with environmental issues. The MoEF itself has restricted its engagement with only a few ENGOs and the ones with which it engages may be influenced by foreign entities. This is the real danger to guard against by policy makers of any developing country.

Strategy of Environment

It is adequately clear that by the 21st century, the entire focus of attention has shifted to climate change and its security implications in the West. In the US, there are many NGOs that conduct research in this area; some of them include the American Security Project (whose projects include Environmental Security's Climate Change and Military Project), Bipartisan

Policy Centre, also known as the National Commission on Energy Policy (with three main areas of study—workforce preparedness for future energy jobs, regional climate change impacts and adaptation planning, and climate policy impacts on energy-intensive manufacturing), the Brookings Institution (focussing on energy security, economics and environment), the CNA Corporation (that runs a project on National Security and Climate Change), Centre for New American Security (CNAS, that has a project on Energy Security and Climate Change), Centre for Strategic and International Studies (whose Energy and Climate Change programme also includes the Asian Regionalism Initiative, which explores the ways in which the Asia-Pacific can work together to address the challenges of energy insecurity, climate change, and humanitarian crises), Council on Foreign Relations (that released a report in 2007—'Climate Change and National Security: An Agenda for Action'), National Security Network, to name just a few. When it comes to the other Western countries, some of the prominent ones comprise Adelphi Research (that has undertaken more than 50 projects in the EU, the US and developing and transition countries on international environmental policy, European environmental policy and policy integration, climate change and institutional aspects of environment, conflict and cooperation), Advisory Council on Global Change (WBGU, that was created by the German federal Government), Chatham House (a London-based organisation that has one of its research areas as the Energy, Environment and Development Programme), E3G (otherwise known as Third Generation Environmentalism which is an independent, non-profit European organisation), the Institute for Environmental Security (whose programmes include Climate Change and International Security, Climate Change and the Military, Environmental Security for Poverty Alleviation, the FUEL project to integrate energy needs in humanitarian crisis situations, Greening European Society, Global Policy Coherence, the Hague Environmental Law Facility, the Inventory of Environment and Security Policies and Practices, and the Pathfinder Programme to restrict the import of illegally extracted resources from conflict zones), Royal United Services Institute (RUSI, that is engaged in cutting edge defence and security research), and so on.[35] Out of these, the majority is funded heavily by the

Government as well as private agencies alike. The involvement of policy-makers in these bodies is also well-known. For example, take the case of the CNAS, a research institution where Sharon Burke was the Vice President for National Security; she concentrated on national security implications of global natural resources challenges and climate change; in 2009, she was nominated by President Barack Obama to the post of Director of Operational Energy Plans and Programmes at the Department of Defense.[36] This is the kind of coordination that is required in India between the Government and the research institutes/think tanks. Another example would be the Bipartisan Policy Centre, which was established in 2007 by former senate majority Leaders Howard Baker, Tom Dasche, Bob Dole and George Mitchell.[37] This is another form of coordination between the decision-making bodies and the research and development agencies, wherein the current or former policy-makers themselves take the plunge into research activities to generate policy proposals for the establishment. Apart from these, the CNA Corporation is a federally funded research organisation that operates the Center for Naval Analyses and the Institute for Public Research.[38] This is a classic example of how the Government itself takes complete control of a research entity in order to carry out bipartisan study on issues of national security. In India, the main challenge in terms of engaging with the NGOs and the ENGOs is not the lack of resources but political will, rather 'bureaucratic will'.

It is important to note that all these NGOs/think tanks/research institutes are not dedicated to environmental research. In a highly-interconnected world, in which every issue is connected to every other issue, it is totally ludicrous to have such organisations with single focus. Take the case of RAND Corporation that was established with the mission of connecting military planning with research and development decisions in the US. Its core research areas now include: Children and Families, Law and Business, Education and the Arts, National Security, Energy and Environment, Population and Aging, Health and Health Care, Public Safety, Infrastructure and Transportation, Science and Technology, International Affairs and Terrorism and Homeland Security.[39] In India, organisations such as TERI and DA are wholly devoted to environment-related research. The defence think tanks deal mostly with the traditional national security threats. There are a few think tanks that have been carrying out multi-faceted research including environmental security such as Institute for Defence Studies and Analyses, Centre for Air Power Studies, Centre for Policy Research etc. However, neither the quantum of research nor the amount of human resources doing research on environment is adequate besides the fact that the Government hardly pays heed to the work being done in these organisations. This is the reason why the fresh researchers are completely dependent on Western literature on this subject which makes their understanding of it highly skewed and biased. The need of the hour is to form the Indian perspective on the subject and formulate policies that would be in India's interest. The fact of the matter is environment cannot be segregated as it is clearly part of every nation state's strategy. The West has realised this and is using its organisations even to influence policy-making in other countries by accomplishing rigorous environmental diplomacy. In fact, since most of these organisations are non-profit and independent, it is easier for them to build a global network by which they could interact with their counterparts in developing countries where this culture is absent or negligible. These networks could easily come in handy at the negotiations too with the Track II diplomacy shaping Track I diplomacy. With their presence cutting across political boundaries, they could even become ambassadors of their respective countries, thus projecting a very favourable image of their countries.

India is far from achieving the Western standards. India is yet to recognise how environment can be a part of its strategy at the regional and international levels. The West started by treating the NGOs as part of a 'collective bargaining' formula; now they have progressed to a stage in which environmental issues are being exploited by them to influence decision-making all over the world. However, in India's case if the right decisions are made at the right time, the situation could change. India should follow the Western model which advocates the use of think tanks to carry out environment-related research as well as an extension of its foreign policy. Just as there is a need for coordination among various agencies, there is also a need for having more think tanks with wider agendas and not restricted to one or two issues. The dividing line between the ENGOs and the NGOs has to be erased so that the research conducted in this area is much more comprehensive, strategic and could be used to enhance India's influence elsewhere in the world. The divide between the think tanks and the Government also has to be bridged. India has to identify its target countries to establish its Track II diplomatic efforts in the field of environment. This has already been discussed in the fourth chapter. The various ministries dealing with environment have to converge besides the think tanks which are a potent knowledge base.

Culture: Progressive or Regressive? A Challenge

The reason for diplomatic failures on the part of the Government, in terms of engaging with the public, cannot be lack of resources or expertise alone as it does not require high-end technology. The answer is simple: culture. It is more important to realise that the need of the hour is not only to occupy the moral high-ground but to go one step ahead after the constructive results are achieved—to throw light on them. Culture has to be dynamic and progressive; disregard of diplomacy of any sort can only harm the cultural ethos of a country. In the similar vein, the Government will end up harming themselves by neglecting environmental diplomacy. This is one subject that has the potential to immediately attract public attention as they are able to relate to the subject due to an information blitzkrieg on the subject already unleashed by the media—both, print and electronic.

The benefits of environmental diplomacy are numerous. It has to be kept in mind that diplomatic initiatives serve the interests of the Government, the ENGOs as well as the general public. First, if the Government takes interest in launching such initiatives, they will reflect the efficient functioning of the Government which would supplement its public diplomacy strategy besides enhancing credibility and integrity. Second, the emergent image of the Government as a cohesive unit is expected to give confidence to the people about the nation's progress. The all-important 'connect' between the organs of the Government and the people could help the former firmly hold the reins of power. Third, it would motivate the people to take up responsibility to work on environmental issues. Environment is such an all-encompassing subject that the coordination between the different Government agencies as well as between them and the ENGOs is crucial. Similarly, implementation of the Government policies at the grassroots level would be facilitated by healthier public participation. It is a time-tested fact that any policy of the Government, when implemented, can reap benefits in a sustained manner only if the public is kept attuned of these policies as well as involved in a full-fledged manner; this would ensure that the 'fruits' of the policy could be 'consumed' in future too irrespective of the various uncertainties associated with the establishment.[40] This would also guarantee that instead of creating friction between the Government and the ENGOs, bridges would be built that would be used by the public for communication.

The Meeting Point of Different Strands of Thought

It is very clear that the role of environmentalism has been immense since time immemorial. The traditionalists would argue that their role has largely been restricted to creating awareness (both positive and negative) among the masses and putting pressure on the Government from time to time; and that the systemic change that the ENGOs have been demanding for years has not been realised despite the impending environmental catastrophe. Thus, on the one hand, environmentalism has indeed played a significant role in propelling critical environmental issues such as global warming and climate change. On the other hand, the impact of the ENGOs on environmental policy formulation is minimal, especially in a developing country like India. Others would emphasise on the point that the Government has been actively involving the ENGOs in policy implementation due to reasons ranging from lack of resources and expertise to a conscious effort to increase public participation in nation-building. The other reason why the Government would like to join hands with them is to ward off criticism and protect the vote-bank. An inefficient Government may be susceptible to the looming influence of the ENGOs.

The evolution of environmentalism and the emergence of various environmental organisations have led to multi-layered governance. This has been possible despite the stronghold of the State because of the indivisible nature of environment and the pervasive plus destructive nature of environmental problems. From radical to modern to new environmentalism, every stage has augmented the potency of this movement in effecting social and political change. One cannot deny the fact that the ENGOs have turned into political actors, either attempting to represent the citizens who are marginalised by the state system or influencing policy formulation in public interest. The strategies used by the ENGOs have been dynamic. From distribution of pamphlets and production of literature before the advent of information revolution, to the use of technologies by the current generation environmentalists, environmentalism has come a long way.

The role of environmentalism and the ENGOs in a liberal democracy is expected to increase in the coming years especially since both compliment ideas of freedom and justice. The concept of representation could bear fruit in a system where democratisation of every unit of governance is realised. The question of who represents who has been partially resolved by the States by opening the doors of governance to actors who have the resources to manage local communities as well as local environment. However, a few sacred doors are yet to be opened that could mainstream environmentalism in national policy formulation that has not taken off in many countries and is progressing at a very slow pace in others. The monopoly of the States over climate change policy is stringent to such an extent that not a single international agreement has been successful since the beginning of the negotiations. The possibility of governance (environmental governance) without a Government is thus next to impossible. As one can see from the various examples mentioned in the chapter, there are instances where the Governments have clearly diffused some of its responsibilities to the NGOs, empowering them to a certain extent. Some of the NGOs have been integrated with the formal system that essentially erases the line dividing the State and the larger civil society as well as negates the independent character of these NGOs. At the end of the day, one could also suggest that some of the NGOs that begin by representing the citizens end up representing the States in order to be heard.

Diffusion of governance is best exemplified by the so-called partnerships between Governments and the NGOs. Very often national governments and international financial institutions (IFIs) such as the Global Environment Facility direct money to local environmental projects via the NGOs. Policy-makers could take the assistance of the NGOs in effectively implementing environmental policies if they feel that the NGOs are more equipped and experienced in dealing with the local communities. International donors sometimes prefer the NGOs to governmental organisations as the latter might not have adequate infrastructure and resources to implement the projects besides the fact that in a country like India, the will to entertain such projects is also low-key. However, policy-makers have also been accused of using the ENGOs to sell their policies to local communities which they would not be able to do themselves or through governmental organisations as the NGOs are believed to have closer ties with the local communities. The ENGOs have also been charged with accusations of being agents of capitalism as they thrive on the principles of liberalism based on free market economy. This accusation cannot be legitimised against an ENGO; it is more appropriate in the case of a developmental NGO. However, a few ENGOs such as the WWF would be more in the line of fire as they work in tandem with the Governments and the IFIs, while certain others such as the FoE have openly declared their resistance to a neoliberal economic world order.

The global civic society is also a nascent concept since there are several countries that are still not part of the global decision making process. In a world where 99 per cent of the decisions are taken by practically less than 20 countries, the global civic society would take eons to materialize. Although the concept of ecological citizenship could give new dimensions to citizenship, still States would continue to be the dominant factor in the international arena. In global environmental governance, the need for the States to acknowledge and empower the ENGOs is expected to gain salience in the years to come. With the free flow of information and close scrutiny of government policies at every stage, the ENGOs would be in a better position to challenge the States' authority. For this, the legitimacy of the ENGOs itself has to remain intact. One can conclude that the urgency for increased collaboration between the scientists, policy-makers, the ENGOs and the civil society could pave the way for a number of positive outcomes – mitigation of the effects of environmental change both at the global and the domestic levels.

Finally, economics and strategy of environment is equally as important as any other issue. The ENGOs in India are short on resources and thinking. The Government has also contributed to the parochialism that pervades research institutes (both governmental and nongovernmental) in the country by not giving them enough voice that is heard in the corridors of power, resources and space to function independently. In the West, the efforts of these organisations are completely backed by the Government. There need not be any distinction between the NGOs and the ENGOs. The Indian NGOs and think tanks should now look forward to using environment as a strategy in regions such as South Asia, Southeast Asia, Africa and Latin America. International relations are all about give-and-take in the 21st century; the world is no longer unipolar. Therefore, unless India takes proactive steps to engage at different levels with the countries in the above-mentioned regions, it can never be a great power.

Notes

1. Deba Prashad Chatterjee, "Sustainable Development: Environmentalism versus Appropriate Technology," in R. K Gupta, Klaus Bosselmann and Prasenjit Maiti (eds.), *Global Environment: Problems and Policies* (New Delhi: Atlantic Publishers & Distributors (P) Ltd., 2008), vol. 4, p. 9.
2. Bron Taylor, "Radical Environmentalism," *College of Liberal Arts and Sciences Publication* (Florida: University of Florida), p. 1326, see http://www.clas.ufl.edu/users/bron/ern/R.pdf, accessed on December 15, 2011.
3. n. 1, pp. 12-14.
4. Jeff Walker, "The New Environmentalism," *Harris/Decima* (Ottawa: Canada Post), p. 2, see http://www.the-cma.org/PublicUploads/227526TheNewEnvironmentalism.pdf, accessed on December 16, 2011.
5. Neil Carter, *The Politics of Environment (Second Edition)* (New York: Cambridge University Press, 2007), p. 149-150.
6. Laurie A. Kutner, "Environmentalism Activism and the Internet," in R. K Gupta, Klaus Bosselmann and Prasenjit Maiti (eds.), *Global Environment: Problems and Policies* (New Delhi: Atlantic Publishers & Distributors (P) Ltd., 2008), vol. 1, p. 184.
7. Ibid., p. 186.
8 Klaus Bosselmann, "The Earth Charter and Global Environmental Governance," in R. K Gupta, Klaus Bosselmann and Prasenjit Maiti (eds.), *Global Environment: Problems and Policies* (New Delhi: Atlantic Publishers & Distributors (P) Ltd., 2008), vol. 2, pp. 65-67.
9 Ibid., p. 69.
10. For more information, see http://www.earthcharterinaction.org/content/pages/

Read-the-Charter.html, accessed on January 04, 2011.

11. n. 8, p. 78.
12. n. 8; p. 80.
13. Derek R. Bell, "Sustainability through Democratisation? Assessing the Rule of Environmental NGOs in a Liberal Democracy," *NGOs, Sustainability and Democracy* (New Castle: Political Studies Association, University of Newcastle), p. 1, see http://www.psa.ac.uk/journals/pdf/5/2003/Derek%20Bell.pdf, accessed on December 15, 2011.
14. Ibid.
15. Ibid., p. 6.
16. Ibid., p. 12.
17. Earth Science and Climate Change, see for more information, http://www.teriin.org/index.php?option=com_division&task=view_div&id=26, accessed on December 15, 2011.
18. "Interim Report of the Expert Group on Low Carbon Strategies for Inclusive Growth," *Planning Commission, Government of India* (May, 2011), see http://planningcommission.nic.in/reports/genrep/Inter_Exp.pdf, accessed on June 20, 2011.
19. "The ENVIS CENTRE at Development Alternatives," Development Alternatives Information Network, see http://www.dainet.org/envis.htm, accessed on January 08, 2012.
20. See for more information, "ICC and the Business and Industry Coordination Process at UNFCCC," http://www.iccwbo.org/policy/environment/unfccc/cop15/id33610/index.html, accessed on January 14, 2011.
21. Fiona Godlee, "Rio Diary: A Fortnight At The Earth Summit," *British Medical Journal* (Britian: July 11, 1992), vol. 305, no. 6845, p. 103, see http://www.jstor.org/stable/29716265, accessed on January 04, 2012.
22. Elisabeth Corell and Michele M. Betsill, "A Comparative Look at NGO Influence in International Environmental Negotiations: Desertification and Climate Change," *Global Environmental Politics* (Massachusetts Institute of Technology, November 2001), vol. 1, no. 4, p. 90, see http://www.colostate.edu/dept/PoliSci/fac/mb/Comparative%20Look.pdf, accessed on December 15, 2011.
23. Ibid., p. 97.
24. System Change – Not Climate Change: A People's Declaration from Klimaforum09, December 2009, see for more information, http://old.klimaforum09.org/IMG/pdf/A_People_s_Declaration_from_Klimaforum09_-_ultimate_version.pdf, accessed on December 14, 2011.
25. "Copenhagen Climate Change Conference: John Vidal Goes Behind the Scenes," *The Guardian* (December 08, 2009),see http://www.guardian.co.uk/environment/

video/2009/dec/08/copenhagen-climate-change, accessed on January 07, 2012.

26. David B. Ottaway and Joe Stephens, "Inside the Nature Conservancy - Washington Post Expose," *Washington Post* (May 04, 2003), see http://lists.envirolink.org/pipermail/ar-news/2003-May/000054.html, accessed on January 04, 2012.
27. See for more information, "Brent Spar Greenpeace Vs Shell," http://wn.com/Brent_Spar_Greenpeace_vs_Shell, accessed on January 07, 2012.
28. n. 5, p. 151.
29. "Enabling IT Innovations," Cool IT, see http://www.greenpeace.org/india/en/What-We-Do/Stop-Climate-Change/Green-Electronics/, accessed on January 05, 2012.
30. Planet Heritage Foundation, Inc., "Climate Change and National Security: A Field Map and Analysis of Funding Opportunities," July 01, 2010, pp. 9-10, see http://www.sherryconsulting.com/Final%20Climate%20Report.pdf, accessed on January 17, 2012.
31. Ibid., p. 7.
32. See for more information, http://www.ashanet.org/munich/fcra.html, accessed on January 17, 2012.
33. Manoj Kumar and Dhanasree Jayaram, "Diplomacy and Military: Talk the Walk," *Defence and Diplomacy* (New Delhi: Centre for Air Power Studies, October, 2011), vol. 1, no. 1, pp. 54-56.
34. Interaction with Dr. Lavanya Rajamani, Professor, Centre for Policy Research (CPR), at CPR (New Delhi), on July 11, 2011.
35. n. 30, pp. 14-27.
36. n. 30, p. 17.
37. n. 30, p. 15.
38. n. 30, p. 16.
39. For more information, see http://www.rand.org/research_areas.html, accessed on January 17, 2012.
40. n. 33, pp. 56-58.

7

Environmental Change and the Fabric of National Security

"To be on the alert is to live; to be lulled into security is to die."

— Oscar Wilde, Irish Writer and Poet

The interconnectedness between environmental change and national security is one of the most debated and controversial subjects in international political, security and strategic studies. It is contentious not only because of the general human tendency to live in the present 'completely' but also due to politically and strategically motivated research on the subject. At the outset, it seems outrageous to disconnect environment from security considering the rising number of people being killed in natural disasters in many parts of the world or the lowering water tables that have led to extreme droughts and famines in other parts of the world. Quite contrary to this belief, the combination of national security and international politics has transformed the obvious into a dubious, endless debate. In the present scenario, it would be far-fetched to supersede sovereignty and establish global governance that disregards political boundaries. Therefore, it is imperative for the research community to go beyond the socio-economic realm of environment and analyse its geopolitical and geostrategic dimensions that are expected to dictate international relations in the coming years to a great extent. Countries have begun to use environment as a strategy rather than promoting the idea of global commons. Environmental policies and security strategies have gone global and hence, it is imperative for India to revaluate its stated position on environmental change and national security if it wishes to have a say in global decision-making. Whether it is national security or international security, keeping environment out of the equations may not be a bright idea for any country for two reasons – first, the real security concerns that could emerge in future and second, because of the existing misperceptions about the issue that could percolate into the decision-making bodies due to imperfect systems of governance leading to formulation of disastrous policies and strategies. While looking at the security paradigm, it becomes crucial to explore the sustainable development paradigm too in order to throw light on the Indian perspective on the issue.

The chapter, to begin with, makes an attempt at dissecting different threads of national security and how it is perceived by the international community at different levels. It scrutinizes the ways in which environmental change could jeopardise national security by analysing the issue from the perspective of both the state and the individual as well as through the lenses of international relations (IR) theories such as neo-realism and neo-liberalism. A clear-cut study of Western literature on security implications of climate change in order to establish its pitfalls has been supplemented by an evaluation of the role of the UN in dealing with climate change especially in securitising it, which has been opposed by India. The sustainable development paradigm would be briefly touched upon as it forms the foundation of the benchmarks for understanding the Indian perspective on the issue. Seeds of realignment of India's strategy at international parleys and negotiations (rather than mere revisionism that is confrontational and likely to be suppressed systematically by the opponents) are what the country's leadership needs to sow for its diplomatic efforts to bear fruit at the international level as well as address environmental change at the domestic level.

National Security and Environmental Change: A Theoretical Perspective

State-Centric Approach towards National Security

One of the views associate the concept of national security directly and blatantly with the Peace of Westphalia which brought to the fore the concept of a 'sovereign state'. When national security is defined from the state's perspective, its entire notion is linked to the requirement to maintain survival of the state through various means. In this case, even if the definition encompasses protection of citizens, the term 'citizen' itself connotes the existence of the state. In fact, both realists and neo-realists have agreed that it is the 'security dilemma' which is a source of conflicts between states. Walter Lippmann says, "A nation is secure to the extent to which it is not in danger of having to sacrifice core values if it wishes to avoid war, and is able, if challenged, to maintain them by victory in such a war."[1] Neo-realists (offensive) such as Mearsheimer go one step ahead to assert, "Cooperation among states is constrained by the dominating logic of security competition, which no cooperation can eliminate."[2] The realist perspective clearly defines national security in such a manner that it could be endangered only by external factors.

Liberal institutionalism on the other hand also recognises the problem of security dilemma, yet advocates of this theory believe that international institutions can not only avert the dangers of security competition through cooperation and complex interdependence but also achieve international security. According to Keohane and Martin, "Institutions can provide information, reduce transaction costs, make commitments more credible, establish focal points for coordination and in general, facilitate the operation of reciprocity."[3] Still the focus is on the state since the term 'international' connotes political boundaries and it acknowledges the existence of nation states. Yet another definition of national security was given by Harold Brown, the US Secretary of Defence from 1977 to 1981 in the Carter administration, which is, "National security then is the ability to preserve the nation's physical integrity and territory; to maintain its economic relations with the rest of the world on reasonable terms; to preserve its nature, institution, and governance from disruption from outside; and to control its borders."[4]

This is where distinction is drawn between the traditional and non-traditional security threats (both being state-centric), wherein issues such as environmental change usually fall under the latter's category. At the same time, some analysts feel that environmental change is very much a traditional threat. According to Cleo Paskal, classical strategists such as Kautilya and Sun Tzu very clearly articulated how the terrain, time, supply lines and leadership were crucial components that had to be kept in mind during military confrontations or engagements. Time has come when rapid environmental change has started to impact all these factors; this in turn brings the whole international community to acknowledge Kautilya's words that emphasised that power, time and space are mutually dependent and not exclusive. This brings to attention the fact that environmental change could impact strategic planning and analyses thereby making it very much a traditional threat. She remarks, "When it comes to assessing environmental change, it is time to go back to tradition."[5] In contrary to this belief, others have argued that environment, poverty, natural disasters and so on should not be treated as non-traditional security threats as this could give non-democratic regimes an opportunity to 'expand into areas it had not operated in previously'.[6] This view essentially suggests that rather than viewing environmental change from a state-centric point of view, which could give autocratic countries such as China to gain control over virtually every aspect of human life, it is advisable to include environmental change in the human security paradigm that looks at every concept from an individual-centric point of view. It is one thing that the state is increasingly being dragged into a myriad of issues that have repercussions for the existence of the state itself (like in the case of island nations that might lose statehood if their territories are lost to environmental change) but it is a totally different thing to non-traditionalise an issue to encroach upon the rights of the individuals in the name of providing them security. There is yet another view that human safety should be treated as different from human security in order to limit the purview of security architecture as human safety is protection against inadvertent events while human security is protection against intentional damages. Since it has been largely accepted that environmental change is a reality and it could well be caused by human activities in most of the cases, it becomes a human security issue as unresponsiveness on the part of the Government to such dangers is equally intentional as a deliberate attempt to use environment as a tool to harm an individual's interests.

Individual-Centric Approach towards National Security

National security could be defined in such a manner that 'individual' instead of state forms its core. That essentially means human security although some equate human security directly with non-traditional security threats as both cover similar areas. Take the case of India. The Preamble of the Indian

Constitution states that India is a 'sovereign, socialist, secular, democratic republic', and that it strives to secure to all citizens of India – 'social, economic and political justice'; 'liberty of thought, expression, belief, faith and worship'; 'equality of status and opportunity', and 'to promote among them all fraternity assuring dignity of the individual, and unity and integrity of the Nation'.[7] Very clearly, the Indian Constitution asserts the democratic principle that power rests in the hands of the people. According to Air Commodore (Retd) Jasjit Singh, national security is nothing but the preservation of the core values of a nation that are prescribed in the Preamble (that is, justice, liberty, equality and fraternity). Secondly, it also entails the protection and promotion of the vital national interests.[8] National interest is nothing but the 'general long-term and continuing purpose which the state, the nation, and the government all see themselves as serving'.[9] Protection of environment is definitely in India's national interest. Each of these concepts can be directly linked to security. John Rawls elaborates on the conception of justice in the form of two principles: "First, each person participating in a practice, or affected by it, has an equal right to most extensive liberty compatible with a like liberty for all; and second, inequalities are arbitrary unless it is unreasonable to expect that they will work out for everyone's advantage, and provided the positions and offices to which they attach, or from which they may be gained, are open to all. These ideas or principles express justice as a complex of three ideas: liberty, equality, and reward for services contributing to the common good."[10]

Justice encompasses every other concept that is mentioned in the Preamble. Therefore, every country is supposed to formulate policies that cater to the 'common good', failing which the system of governance would also fail, causing various types of disruptions—political, social, economic and environmental. One factor that strikes the reader is that the Preamble assures 'justice' but the ambit is restricted to political, economic and social facets. Environment, which forms the base of all social, political and economic relations is left out, which is the case in nearly all countries. The US Environmental Protection Agency defines 'environmental justice' as "the fair treatment and meaningful involvement of all people regardless of race, colour, national origin, or income with respect to the development, implementation, and enforcement of environmental laws, regulations, and policies."[11]

Even in India that has a huge environmental heritage, there have been numerous cases of violations of environmental justice especially against the tribal communities whose homelands have been destroyed due to deforestation or unequal distribution of food and energy across the country. This is also breach of national security that is generally overlooked by the Government as a security concern. If the definition prescribed by Air Commodore (Retd) Jasjit Singh is taken into consideration, then the issue of governance would form the nucleus of national security as the political system is directly accountable to the people and without the people's consent, the political system would not exist. Since this chapter would like to restrict itself to the environmental arena, environmental governance would gain relevance while analysing environmental security from the state's perspective. Environmental governance is one of the elements of overall governance. Any country's environmental governance does not stop at the duties and decision-making of the policy-makers; it extends to different levels – individual, community, State Government, Central Government and agencies, and international treaties. In this closely-knit system, judiciary plays an important balancing role at all the levels.[12] Therefore, national security covers every single atom that exists in a system and it could be achieved only if all of them are protected from both internal and external threats. National security includes all-embracing activities that focus on 'providing the right to life and liberty to its citizens; equality in all spheres of activity; cooperation between the public and private sectors operating in the state; preservation of territorial sovereignty and integrity; maintaining a flexible civil and military relation; robust economic development and a proactive and independent media'.[13]

Environment and Theories of International Relations

One of the important strands of national security, as discussed above, is environmental security. However, this chapter does not strive to look at everything related to environmental security. The inclination is more towards the impact of environmental change on national security. In a narrow sense, environment security is usually referred to the relationship between security concerns such as armed conflict and the natural environment. This is the reason why the focus has principally been on the environmental damage caused by armed conflicts or environmentally-caused conflicts. Currently, transnational environmental problems have also been added to the ambit of environmental security; these include climate change, deforestation and loss of biodiversity. It could be argued that environmental security should now encompass other forms of security as well such as energy, economic, political and social. It is understood that the effects of climate change would have direct as well as indirect bearing on a nation state's energy security with increased pressure on countries to adopt renewable energy technologies whose unpredictability factor is considered to be greater than traditional sources of energy such as fossil fuels. Besides, extreme weather events attributed to climate change itself can affect the existing energy infrastructure affecting the energy security of the nation. Close linkages between climate change and economics have already been established by the climate change negotiations. The costs of adaptation and mitigation are being evaluated by every nation state with fears of these costs overriding the benefits prevalent all across the international community. The global carbon market is yet another paradigm of the thriving economics of climate change. Political and social security is also threatened by environmental change. For example, it could take away millions of livelihoods leading to unemployment. The unemployed population could indulge in anti-social activities leading to increasing levels of crime. All these aspects have already been discussed elaborately in the previous chapters.

When environmental change is seen through the prism of realism and neo-realism; one could say that such a phenomenon could lead to further conflicts between nation states rather than cooperation. One could even illustrate this reality by giving the example of the global climate change negotiations steered by the UN that started with the ambition of designing an international framework to address the growing challenges emanating from the changes in climate, but have instead turned into a platform of conflicting ideologies and interests. The interests of the States overrule the needs of global environmental governance. Even the Rio Declaration of

1992 plainly defends state sovereignty over natural resources. The second commitment period of the Kyoto Protocol hung in the balance until the Durban Summit as developed countries feared that developing countries especially India and China would gain advantages over them. Besides, within developed countries, the EU is sceptical of the commitments of the Umbrella Group towards meeting the targets set by the Kyoto Protocol. According to the EU, these countries could violate the agreement which suffices with the 'problem of cheating'—a primary element of realism and neo-realism. Canada is set to become the only country to have completely undermined the Kyoto Protocol despite both signing as well as ratifying it as it is far from fulfilling its requirements under the first commitment period (2008-2012). It withdrew from it just after the Durban Summit after the nation states came to consensus regarding the drafting of the Durban Platform for Enhanced Action and certain developed countries such as that of the EU agreed to enforce the second commitment period of the Kyoto Protocol. The US signed the Protocol but has still not ratified it. These countries have essentially refused to abide by the requirements of the Protocol as they consider it a threat to their economic and energy security that are critical elements of 'national security'. In a way, their apprehensions could be understood in the light of the 'problem of relative gains'—another important aspect that guides realist and neo-realist theories. It is well-acknowledged that if nation states 'cooperate' on an issue such as climate change in an equitable fashion, each and every state is expected to benefit, which is nothing but 'absolute gains'. However, states tend to be concerned with 'relative gains'. Take the case of the recent summits at Copenhagen and Cancun. One of the causes of the 'failure' of these two summits to arrive at an 'international agreement', is touted to be global economic meltdown which affected developed countries most and left emerging countries such as India and China largely untouched. Thus, countries like India and China are no longer regarded by a large part of the international community as developing countries, which posed no economic competition to the West; they have emerged to give serious competition to the West in every part of the world. Moreover, the trust deficit between the developed and developing countries has only augmented over the years. In an increasingly competitive and mistrustful environment, cooperation has given way to conflict.

Environmental Change and Security: Perceptions and Misperceptions

A closer analysis of the Western literature on 'climate change and national security' unravels the geopolitics of environmental change that needs to be not only revised but also overturned to accommodate the interests of developing countries and the Least Developed Countries (LDCs). Some of the contentious claims include:

- The 'regional hotspots' have been identified as North Africa, Sahel zone, Southern Africa, Central Asia, India, Pakistan and Bangladesh, China, Caribbean and the Gulf of Mexico and Andean region and Amazonia. Obvious omissions of the US, Australia, New Zealand, Japan and many other European countries that were hit by natural disasters and unusual weather patterns that have resulted in immense damage, raises questions about the credibility of Western interpretation of environmental security.
- The 'unstable multi-polarity' is being touted as one of the major reasons for 'political, institutional and socio-economic turbulence'. This again points fingers towards emerging players such as India and China that have been blamed from time to time for the inability of the international community to sign a climate change treaty based on 'shared responsibility'. This assumption has been taken advantage of in order to create fissures between the newly industrialising countries and the poorer developing countries.
- Interestingly, the reports very much deliberate on the issue of trans-boundary migration. The language used yet again raises questions and does not address the core issues related to environmental change. The German Advisory Council on Global Change (WBGU) report blatantly observes, "In developing countries in particular, increase in drought, soil degradation and growing water scarcity in combination with high population growth, unstable institutions, poverty or a high level of dependency on agriculture means that there is particularly significant risk of environmental migration occurring and increasingly in scale." It

continues, "Trans-boundary environmental migration will mainly take the form of south-south migration, but Europe and North America must also expect substantially increased migratory pressure from regions most at risk from climate change."[14] This clearly draws a line between how migration could affect developing countries but not developed countries as much as the former, which again is debatable. On the one hand, the West is doing its best to curb immigration due to the ongoing economic recession as well as its 'multicultural dilemma'. On the other, history has shown that conflict zones are not limited to developing countries.

- The West has completely amalgamated climate change related security threats with its national security and national defence strategies. One of the US' reports (that of the CNA) states, "The US may be drawn more frequently into these situations, either alone or with allies, to help provide stability before conditions worsen and are exploited by extremists. The US may also be called upon to undertake stability and reconstruction efforts once a conflict has begun, to avert further disaster and reconstitute a stable environment."[15] This calls for intervention of external powers in other countries, particularly the developing ones, on account of climate change. This is an extremely astute way of lending legitimacy to foreign interventions in future. Increasing penetration of a country such as China in Africa and Latin America has also forced the West to rethink over its policies and strategies in these continents for its greater involvement.
- The documents and reports also declare that climate change would lead to the emergence of failed states apart from nurturing breeding grounds for terrorist and extremist activities. Such mind-boggling findings have been plainly thwarted by India and many other countries. Failed states as we know currently are ones that have 'failed' to fulfil its security and developmental functions, protection of environment being one of them. Therefore, to consider environmental change a cause of state-collapse would be too far-fetched although it can be regarded as a symptom or consequence of state-collapse. At the same time, rise of anti-social elements due to intensification of poverty, caused by environmental change is highly plausible. These anti-social elements could act against the state.

The United Nations and Climate Change Related Security Implications

In order to scrutinise the UN Security Council (UNSC) meetings on security implications of climate change, it is important to dissect the debate surrounding climate change itself. This has been done adequately in the previous chapters. The tendency of certain countries to connect everything —from hurricanes to droughts to even tsunamis—to climate change is not acceptable to many other countries. Inter-state migration has also been a bone of contention between the developed and developing countries. There are many more issues that one needs to explore to study the fault-lines that exist in this debate.

But before discussing the UNSC meetings, it is equally important to throw some light on the UN General Assembly (UNGA) meeting that took place in 2009. The draft was introduced on behalf of the Pacific Small Island Developing States by Nauru. The main issues that dominated the discussion were migration and resettlement. The impact of environmental change, especially climate change on island nations has been well-documented by the scientific community across the world. Islands could become uninhabitable due to two reasons, either by flooding or by the crunch in the availability of freshwater resources. The threat of submergence is directly linked to sea level rise caused by climate change. The availability of freshwater resources could be endangered due to various reasons, salt water intrusion being one among them. Moreover, island nations may not have the financial and energy resources to import or build and operate desalination units. The political establishments of these nation states have set the alarm bells ringing over the future of their territories and population by raising the issues of human rights and sovereignty at various international forums. Adaptation and mitigation are two legs of any country's climate change policy and these island states are no different. What separates these countries from the other countries in the world is that their domestic as well as international policies are centred on survival and national security. The possibility of their disappearance raises several pertinent questions that have regional and global implications. As deliberated earlier, it has been found that environmental change could redraw the world map as well as create tensions or even opportunities of cooperation between nation states. If an entire island is wiped off the earth due to sea level rise or erosion, the resulting expansion of seas could be a source of conflicts between major powers in the vicinity or even thousands mile away. These are prototypes of rhetoric that have dominated island nations' take on the security implications of climate change. They have also been trying to link it with human security by stressing on survival and not directly jumping to a conflict-situation that could be designed by climate change.

However, not all countries agreed with this proposition. Nicaragua, while speaking on behalf of the 'like-minded group' (Bahrain, Bolivia, Ecuador, Egypt, Iran, Iraq, Kuwait, Lebanon, Libya, Malaysia, Oman, Qatar, Saudi Arabia, the UAE, Yemen and Venezuela) underscored the need for industrialised countries to 'promote sustainable development' and to 'adhere to the principle of common but differentiated responsibility'. Some of the representatives of developing countries such as those of Indonesia, Bahrain, Argentina, Brazil and China asserted that climate change should be discussed at the UNGA and the Economic and Social Council (ECOSOC) and not the UNSC as they would like to discuss it from the point of view of sustainable development and eradication of poverty. This statement was in response to the introduction of the subject in the UNSC in 2007 by the UK. Quite contrary to these countries' stand, developed countries such as Canada, Iceland and the US welcomed the initiative; Switzerland went one step ahead to say that it would have preferred a bigger role for the UNSC in addressing the issue. Japan, New Zealand and Australia, which are surrounded by islands vowed to provide assistance to regional small island states.[16] The resolution in the UNGA was adopted by consensus.

The resolution passed by the UNGA was preceded and succeeded by meetings of the UNSC on the issue. The stand of developing countries at the UNSC could be summed up by quoting the statement made by the Brazilian representative, Piragibe Tarrago, who pointed out convincingly that 'an objective analysis' of the issue could be hampered by tracing conflicts to a single issue such as climate change. He also reiterated the stand of developing countries that development is linked more closely to climate change than security and therefore the former should be tackled first and in this, the role of industrialised countries becomes crucial. He did not hesitate from confessing that scientific evidence had shown that the effects of climate change could only worsen in future and that this could indeed create some amount of economic and social instability. The Representative of Indonesia stated, "Future threats to security posed by climate change must be avoided and he called on all States to adhere to the Rio principles, especially the principle of common but differentiated responsibility, the 'Agenda 21', the Johannesburg Plan of Implementation, the UN Framework Convention on Climate Change and the Kyoto Protocol."At this meeting, the representative of India very categorically stated, "A more immediate and quantifiable threat was from possible conflicts arising out of inadequate resources for development and poverty eradication, as well as competition for energy," and not climate change. At the end of the day, the resolution states, "Some delegations raise(d) doubts regarding council's role on issue, while others, particularly small island states, welcome(d) council's consideration."[17]

When it came to the 2011 UNSC meeting, the resolution was introduced by Germany. This time too, the small island nations made a clarion call to the rest of the international community to deem climate change as 'great a threat as nuclear proliferation or terrorism'. India yet again questioned the legality of discussing 'vulnerabilities and threats induced by climate change' in the UNSC. As per the statement, "Such issues needed a broader approach, anchored in development, adaptive capacity, risk assessment and institutional build-up." Bolivia's representative set the tone for discussion by making it clear that "climate change was a real threat to humanity and Mother Earth but the security implications of climate change should be dealt in a forum where the guilty parties did not have seats for life or the right to veto." The whole debate came down to the question of whether climate change is a 'security threat' or a 'threat multiplier'.[18]

A Closer Analysis of the Statements

There is a need to explore some of these statements in order to understand the politics behind it. First, although many of the phenomena that were discussed by the representatives are not directly linked to climate change and are mostly cases of environmental change, the West's insistence on including them in the debate as causes of climate change was not acceptable to developing countries. Second, the use of phrases like sustainable development and eradication of poverty by developing countries is noteworthy. Even a country such as India stressed on the point that other factors such as unavailability of resources could lead to conflicts and not climate change. As already mentioned, sustainable development encompasses environmental, economic and social well-being. If any one of these variables gets affected, its ramifications would be felt in the whole system this in turn would have a direct bearing on the future. Environment is definitely not a constant, and environmental change has the potential to derail efforts to achieve sustainable development without second thoughts. In fact, even poverty could be exacerbated by environmental change due to loss of physical, energy or psychological infrastructure. To delink anything that affects security of the future generations from the security architecture is ludicrous. Sustainable development is all about addressing the requirements of the future generations without which their survival would be imperilled. This would entail an assessment of the catastrophic impacts of environmental change on essential resources such as water, food, land and so on as any form of shortage in these could severely endanger the future generations. Therefore, environmental change or even climate change is definitely a security issue. Mitigation of environmental change is possible only by following the path of sustainable development or as some would call it 'smart development'. However, the rationale for defining climate change as a security issue in the current scenario is contestable. To define climate change as a potential threat to international peace and security is accurate but to restrict it to inter-state conflicts is too hard to digest for many policy-makers. It was only in the 1990s that the UN slowly began to pay equal amount of attention to conditions within states. There was a popular belief that the UN should demystify the idea of sovereignty and address 'individual, political and civil rights, as well as the right to basic provisions like food, water, health care, and accommodation...The UN reinforced this new perception that pursuing justice for individuals, or ensuring human security, was an aspect of national interest'.[19] It concluded that violation of individuals' rights could also create fissures between states. The system of states remains paramount, yet there was a marked shift in the way international security began to be perceived. Besides, though the line between international and national security is thin, the repercussions of climate change on national security could be more acceptable than on international security for a country like India. International security is a much broader concept that encompasses all nation states and international organisations and it involves protection of the international system and the sovereignty of nation states, thus making it national security in the global context. The term 'security' itself needs to be contextualised and this will be discussed later on in the chapter. Similarly, the mandate of the UNSC should be to discuss every environmental issue that impacts security and not just climate change. Some even allege that the reason behind bringing up the issue in the UNSC in 2007 was to push a deal at Copenhagen by pressurising countries like India and China and the same happened in 2011 for a deal based on 'shared responsibility' at Durban.[20] This is the exact reason why introduction of climate change in the UNSC has not been a welcome step from the perspective of a lot of countries.

Environmental Change and National Security: The Indian Perspective

An Interview with Prodipto Ghosh[21]

Prodipto Ghosh, who is a member of Prime Minister's Council on Climate Change (2011), categorically says that the narrative built by the West on this issue is in reality built on a very weak foundation with baseless evidences of how climate change 'would' cause migration and along with it, social pressures. According to him, a country such as India is being put under pressure to take the issue of climate change more seriously, and to take affirmative actions with stories of how climate change could force Bangladeshis to come to the eastern part of India. The other region that gets lot of attention is the Sub-Saharan Africa. At the same time, the 'inevitable' migration of people from the Pacific islands to Australia and New Zealand is not mentioned at all. Migration is not only on account of climatic reasons;

moreover not all migration is caused by anthropogenic climate change as in the past cyclones have resulted in mass migration from Bangladesh to the Northeast of India. According to him, climate change affects only 1 to 2 per cent of their GDP; in fact they are gaining 16 sq. km. per year on an average. He also described one of his personal experiences in Bangladesh when he came across camped people on the banks of Yamuna River who continued to pay land revenue to the Government despite losing land due to floods as they could settle on new lands being created due to sedimentation. Natural accretion is also attributed to dams. He added that the real security issue is shrinking of the Arctic ice due to which tensions are expected to spring up in the region with overlapping Economic Exclusive Zones (EEZ) and build-up of naval forces although there is some movement towards cooperation on the issue of sharing of seabed resources. India has been an advocate of 'global commons' especially in the context of the Arctic. He went on to argue that as far as the social tensions are concerned, the impact of climate change on India are not going to unveil themselves for another 30 to 40 years. By that time, India will be roughly close to certain countries of Europe in terms of per capita income. He commented that a country like the Netherlands is more vulnerable than Bangladesh. The only difference is that the Netherlands possesses technologies while Bangladesh has not deployed any. In India, the number of deaths from floods and droughts annually has actually come down.

The political agenda and economic advantages have hijacked the negotiations and it has become a replica of the World Trade Organization (WTO) process. Ghosh agreed with the point that the West was clearly trying to tackle the strategic threat of rise of India and China. Climate change has become an excuse for them to apply brakes on the growth process of these two countries. For example, China is expected to have an economy bigger than the US by 2020. Its defence expenditure could surpass that of the US by 2030. Even the EU's primary concerns are India and China, and not the US. The current institutional arrangements such as the Kyoto Protocol are less favourable to the EU, the US, Japan, and so on. This is the reason why they are desperate to change the existing architecture and make it completely in their favour. On the question of the UNSC debate, he asserted that it was objectionable to introduce the issue in the UNSC as it works on consensus; it should rather be discussed at the UNGA, which works on vote. There is no way India could afford to delink economic growth with energy use even if it has a service economy unless it goes back to the Gandhian way. Unless it develops, people's developmental needs cannot be met.

Ghosh stood by the belief that climate change is a global issue, and therefore, other issues related to environmental degradation could be dealt at the local, national and regional levels. He categorically stated that he did not see any point in bringing the issue of environmental change to the UNSC unless it is introduced as a Trojan horse to bring to the limelight the issue of climate change. In India's environmental policy, primacy is given to 'justice'. Indian perception of sustainable development is more socially inclined than environmentally. He articulated his thoughts on whether sustainable development is an achievable goal or not by giving a very optimistic picture of how there has been a sharp decline in death rates and that there was no runaway increase in the global population. The problem lies in the highly resource-intensive lifestyles of industrialised countries and hence, there is a dire requirement to make them less intensive. Not only do these countries need to shift to non-conventional sources of energy but also need to revamp their preferences and behaviour. On India's sustainable development policy, he reiterated the fact that it is ingrained in the Indian ethos. For instance, the practice of recycling is an institutional one in India and this can be further enhanced by technology. The economic signals, according to him, are in the direction of greater sustainability.

From his statements, one could infer that the establishment is not only sceptical but also paranoid about the Western driven debate on security implications of climate change. Perhaps the Government of India is yet to define sustainable development and human security and how inter-related they are. Social and economic layers of sustainable development could be fostered only if the environmental layer is given equal importance. Even when India talks about justice at the international level, one has to remember that any international policy that is based on the principle of equity at the global level should also be applied to the domestic level. If the per capita principle is applied to the GHG emissions while negotiating at the UNFCCC or discussing at the UNSC, it should also be applied to the distribution and use of energy as already discussed earlier in the chapter. The international political discourse has overtaken the real threat of environmental change to human security which is why only traditional security concerns such as that of developments in the Arctic are being taken note of by the establishment. Moreover, much of the scepticism connected to the security angle of climate change could actually be eliminated if it is replaced by environmental change.

Environmental Change and the National Security Apparatus

In India, the current National Security apparatus does not take environmental change into consideration while framing policies. On the one hand, national security itself is considered a holy cow. On the other, human security is yet to find its rightful place in the national security agenda. Brajesh Mishra, former National Security Advisor (NSA), is of the opinion that environmental change affects national security in a broad sense particularly when it affects food production, drinking water, atmosphere, and so on. However, to call it a national security threat would be 'a bit of an exaggeration'. In the interaction, he admitted that it could have impact on the national existence such as in the case of energy requirements if there is shortage as India is not self-reliant. He reinforced the stand that he or the establishment did not have any disregard for the security aspects of climate change or environmental change. This is under the jurisdiction of the Ministry of Environment and Forests (MoEF) that could always coordinate with the NSA to address security concerns related to environment. The NSA does not directly deal with climate change or environmental change. Mishra also pointed out that there was exaggeration on both sides (that is the developed and developing countries). He said, "We do not want to be stampeded into action. We cannot sacrifice the present for the future."[22] Take also for instance the views of the current NSA, Shivshankar Menon, who separated zero-sum challenges such as terrorist threats, espionage and traditional military threats from non-zero sum challenges such as energy, water, maritime security and others. In one of his speeches, he derided the idea of 'reducing' life to a 'matter of security' by using terminologies such as 'human security'. He also commented that 'thinking about climate change as a security issue' might not help the international community 'identify the responses to it'. Another statement that could open Pandora's Box was that 'by strategic we mean of long-term and primarily military significance' and that 'not many of the non-traditional security challenges that we so blithely list these days actually meet the test of this definition'.[23] This brings everyone to the conclusion that there has to be a balance, and also a question of how critical is a country's leadership to the formation and fulfilment of visionary goals.

Unofficially, the Indian establishment has been sceptical of certain claims made by the small island nations that project a Doomsday scenario. It is a fact that coral islands are more vulnerable than volcanic islands as the latter are far above sea level and are not susceptible to sea level rise as is being portrayed by them. Some of the islands of the Pacific and the Indian Oceans could be threatened but the majority of islands in the world are volcanic that are less vulnerable than coral islands. In such a case, India should take cognizance of the situation in the Andaman and Nicobar Islands and the Lakshadweep Islands that are coral islands; these islands are largely untouched by sea level rise according to some in the policy-making bodies but in reality they are as affected as Maldives that is geophysically similar to these islands.[24] Coral islands are 'built of organic material derived from skeletons of corals and numerous other animals and plants associated with corals. Coral islands consist of low land perhaps only a few metres above sea level'[25] while volcanic islands are the remains of undersea volcanoes that far exceed sea level and have coastal cliffs. Officially India has been largely restrained in raising objections against these countries as it does not want to lose their support that could come in handy at multilateral forums. At a Climate Change conference organised by the University of Mumbai and the Liberty Institute, New Delhi and India Study Centre (INSTUCEN), a Swedish scientist, Nils-Axel Morner vociferously argued that neither was the sea level rising nor did carbon dioxide (CO_2) pose any major threat to the climate. He said, "There is no ongoing rise in the sea level at all and since 1970, it fell by about 20 cm and has remained quite stable for the last 30 to 40 years." He assured that all island nations were safe and called the myth of sea level rise as the 'Greatest Lie Ever Told'.[26] Male' Declaration on Human Dimension of Global Climate Change, presented at the international gathering of Small Island Developing States (SIDS) in November, 2007 drew clear linkages between climate change and human rights; this reinforced the point that climate change deprived human beings of their right to live. At the same time, these countries have also been implicated in climate scandals in the past (as revealed in the third chapter), making them highly vulnerable to economic and political pressures. Some of them have been hyping this issue to promote tourism in their countries which is the only financial and economic source for most of them.

The relevance of sustainable development to India's policy-making, albeit social and economic and not environmental, has been adequately explained above. The Brundtland Report released by the UN in 1987 states, "Sustainable development is development that meets the needs of the present without compromising the ability of future generations to meet their own needs." Within this, the concept of 'needs' refers to the 'essential needs of the world's poor, to which overriding priority should be given'; and 'the idea of limitations imposed by the state of technology and social organization on the environment's ability to meet present and future needs'.[27] All these aspects are closely inter-linked with human security. The other important milestone in the discourse of sustainable development is Agenda 21 of 1992. The Rio Earth Summit produced the Agenda 21 document that outlined the need for and the path towards a 'global partnership for sustainable development'. What stands out in Agenda 21 is that it does not restrict itself to the environmental dimension of sustainable development; instead it pays equal attention to socio-economic dimensions such as 'combating poverty', 'protecting and promoting human health conditions', 'children and youth in sustainable development', 'strengthening the role of farmers', and so on.[28] Therefore, it stresses on the importance of integration of all the elements that constitute sustainable development and dealing with them under one umbrella rather than giving one dimension more priority than the others. Agenda 21 places the onus on the UN to channelise the efforts of the entire international community aimed at achieving sustainable development besides calling for transfer of technological and financial resources to developing and least developed countries. On the one hand, sustainable development removes the barriers between growth/development and environment, thereby negating neither. On the other hand, it also makes sure that the vague definition would bring in all the stakeholders on the table for discussion, policy-formulation and policy-implementation. Since Agenda 21 talks about sustainable development worldwide and not in one country, the basic needs of each country have to be defined. Some countries might place eradication or reduction of poverty above everything else while others might demand appropriate technologies to deal with environmental problems. The role of industrialised countries in the North has also been ambiguously defined except their obvious obligation to share resources. The need for these countries to change their lifestyles from consumerist to sustainable, democratisation of institutions (at the international, regional, national and local levels) in all countries and attainment of global equity have been highlighted by developing countries on various occasions in the past. However, these demands were either ignored or struck down.

Sustainable development is, therefore, closely linked to the individual-centric approach towards national security that gives primacy to social justice, equality and democracy which are enshrined in golden letters in the Indian Constitution just as any other Constitution in the world. One of the major hindrances to the achievement of sustainable development at the international level is lack of economic and social justice within and amongst nations. For example, the Bruntdland Report clearly states, "Environmental damage from global consumption falls most severely on the poorest countries and the poorest people, who are least able to protect themselves."[29] This could imply two things – first, poverty has to be tackled before environmental damage, and second, industrialised countries' environmentally destructive acts could be detrimental to developing countries. What one tends to overlook is the fact that equity needs to be achieved within industrialised countries too. For any environmental policy to bear fruit, social and economic inequalities have to be addressed. At the same time, environmental degradation itself could exacerbate the existing social and economic tensions, thereby rendering the task of eradicating poverty and uplifting masses an uphill task. Thus, any Government needs to address these issues holistically.

The Ehrlich equation (I=PCT, where I is the total environmental impact, P is the level of population, C is the per capita consumption and T is the environmental intensity of consumption)[30] reveals a rather grim story that shatters all illusions of sustainable development. The only manner in which environmental degradation can be contained is to decrease consumption. Mere improvements in productive efficiency, energy efficiency, waste management and reduction in pollution through technological means might slow down the pace of environmental degradation but only till a point. The Rebound Effect and the Piggy Principle explained in the second chapter are two such theories that substantiate this scenario. Therefore, without achieving sustainability, resource security would be affected. Resource insecurity in turn would affect national security in the form of both inter-state and intra-state conflicts. Hence, before securitising the climate change issue, an agency such as the UN Commission on Sustainable Development that was established to oversee and advance implementation of Agenda 21 in every country besides providing policy guidance, has to be strengthened.

Conclusion and Recommendations: A Realignment of India's Strategy

It is high time for India to frame policy that could be used to assert its position in a more robust manner at the international level. Poverty is intricately connected to national security. This has been reiterated by not only India and other developing countries but also the West. The problem again lies in the manner in which the two look at the concept of poverty vis-a-vis environmental change. For the West, poverty leads to conflict that in turn entails military intervention. For India, poverty leads to instability that has to be addressed socially and economically. If India could articulate its view of poverty and environmental change in a coherent manner at the international forums, it would be in a better position to negotiate. India should be able to pressurise the West to bring to the table issues of global resource sharing and the thriving inequality. The global institutional system is such that there is no scope for eliminating disparity between the West and the rest. According to the Neo-Marxist Dependency theory, the wealth of the poorer nations tends to decrease as the wealth of the richer nations increases. In such an unequal system, the poor nations are always dependent on the rich nations for their developmental activities. The West has hardly shared any of its resources with the rest of the world despite being responsible for climate change. Unless and until the West agrees to do this, it has no business to tell the rest of the world what it should do to mitigate the effects of climate change that have security implications. The West has to clear its backyard in terms of recognizing the fact that climate change is equally a security threat to North America, Europe and other industrialised parts of the world.

Poverty alleviation is on top of the agenda for the Indian establishment at present. Raising security concerns at this stage is being perceived by developing countries as a ploy by developed countries to overlook poverty eradication and impose binding emissions cuts on the former. In fact, they have even enforced laws under which even trade sanctions could be slapped on countries that miss the targets. Ironically, the US which has not even ratified the Kyoto Protocol passed the American Clean Energy and Security Act in 2009, after introducing the Waxman-Markey Amendment that calls for imposition of trade sanctions on countries that fail to fulfil their commitments after 2020.[31] According to India and other developing countries, climate change could exacerbate the existing poverty which in turn is caused by uneven distribution of resources and wealth in the world. Therefore, priority should be given to eradicating poverty from the society. However, the West has been propagating the idea that climate change could definitely hinder economic development, which is not as much a security concern as climate change induced inter-state conflicts. The West does not wish to fund poverty alleviation programmes in developing countries. In both cases, the debate comes down to hardcore economics which drives international relations. As explained in all the chapters, the international political economy rules over all high politics and low politics.

Figure 1: Understanding the Indian and Western Aproaches

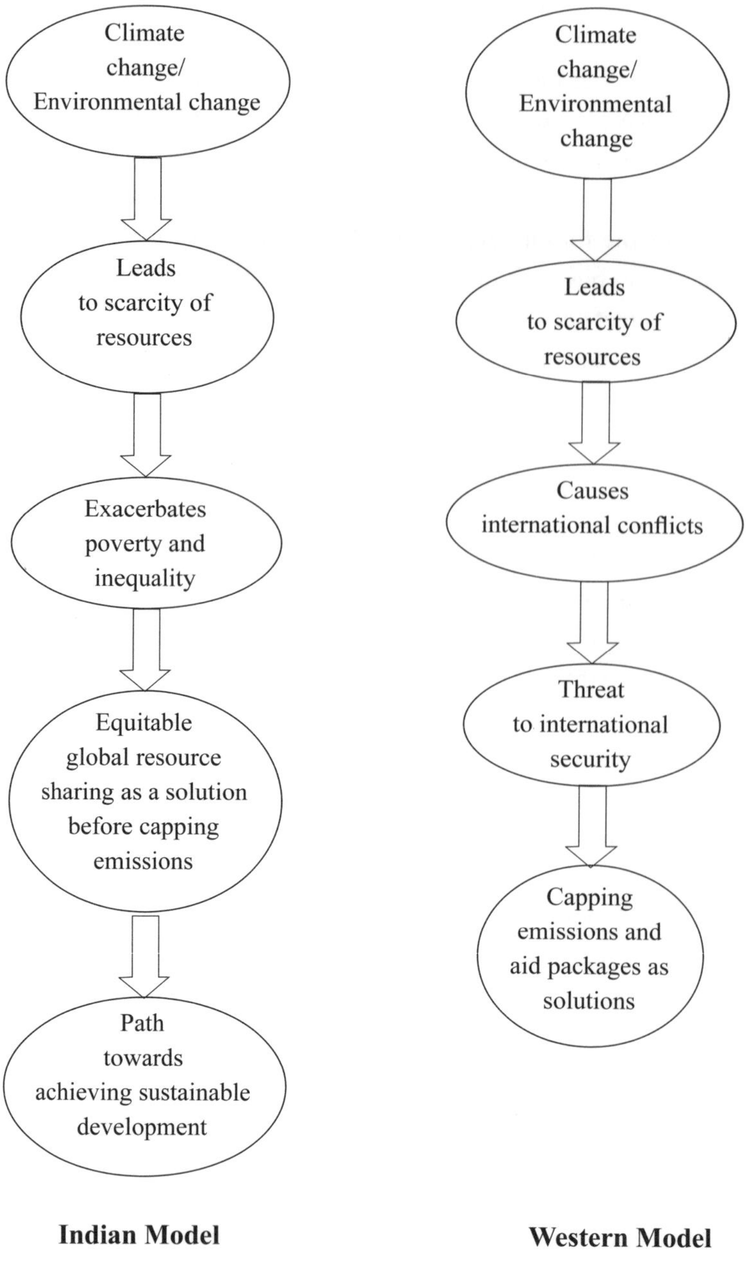

Figure 3: Proposed Model for India's Stand on Environmental Change and National Security

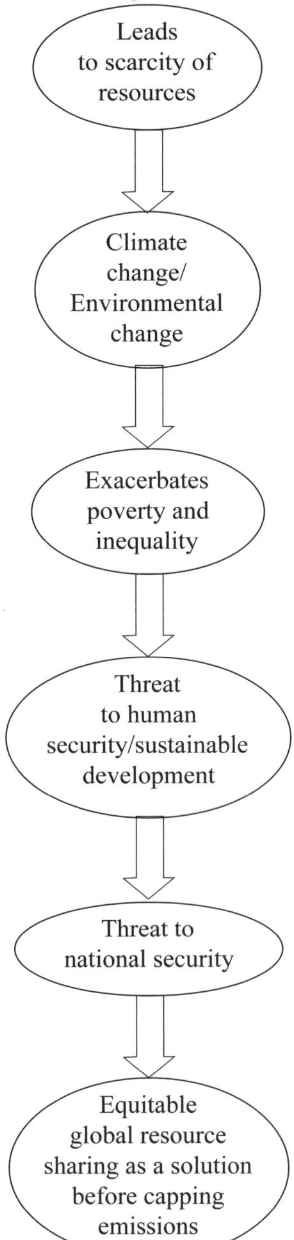

As the proposed model suggests, India only has to add two more layers to its line of thought in order to counter the arguments put forth by the West that are clearly based on their interests. India has to drastically change its strategy when it comes to its interactions with the international community as well as international bodies such as the UN. It is a resentfully accepted fact that all the existing international institutions are a creation of the West and are still dictated by the West albeit things are changing now with the emergence of newly industrialising countries and economic downturn in the West. India and certain other developing countries have been in a 'revisionist' mode to render international institutions fairer and more potent. India needs to free itself from the shackles of the traditional approach to security. The country's policy-makers can join hands with the military and think tanks in order to give more credence to the revised outlook towards security. In the West, the military has taken the initiative to weave strong inter-linkages between climate change and national security. It is time for the Indian military to delineate various aspects of national security that are associated with environmental change. There are certain aspects which come directly under the umbrella of the national security paradigm, water security, safety and security of military installations, possible migration being some of them. It could then advise the MoEF to deal with human security issues that could be triggered by environmental change, food security being one of them. This would help in the formulation of India's national security and strategy policies both at the national as well as international levels.

Notes

1. John Baylis, "International and Global Security," in John Baylis, Steve Smith and Patricia Owens (eds.), *The Globalization of World Politics: An Introduction to International Relations 4e* (New York: Oxford University Press, 2008), p. 229.
2. Ibid p. 231.
3. Ibid, p. 233.
4. See for more information, February Topic Overview, http://debate-central.ncpa.org/february-topic-overview/, accessed on November 20, 2011.
5. Interaction with Cleo Paskal, Associate Fellow, Chatham House, at the 14th Asian Security Conference, New Delhi on February 13, 2012.
6. S. Clifford, "Theory Talk #42: Amitav Acharya on the Relevance of Regions,

ASEAN, and Western IR's False Universalisms," *Theory Talks* (Goteborg, Sweden, August 10, 2011) see http://www.theory-talks.org/2011/08/theory-talk-42.html, accessed on February 22, 2012.

7. M. V. Pylee, *India's Constitution* (Nw Delhi: S. Chand & Company, 2009), p. 33.
8. Interaction with Air Commodore (Retd.) Jasjit Singh, Centre for Air Power Studies, at Manipal University, on November 13, 2011.
9. Urmila Sharma and S. K. Sharma, *Principles and Theory of Political Science* (New Delhi: Atlantic Publishers & Distributors, 2000), p. 126.
10. John Rawls, "Justice as Fairness," see http://www.hist-analytic.org/Rawlsfair.htm, accessed on December 19, 2011.
11. See for more information, Environmental Protection Agency (USA), Environmental Justice, http://www.epa.gov/environmentaljustice/ , accessed on January 15, 2012.
12. Manoj Kumar, *Environment Change and National Security,* (New Delhi: KW Publishers in association with Centre for Air Power Studies, 2011), p. 206.
13. Satish Kumar, "National Security Environment," in Satish Kumar (ed.), *India's National Security Annual Review 2010,* (New Delhi: Routledge, 2010), pp. 14-20.
14. German Advisory Council on Global Change, *Climate Change as a Security Risk* (UK: Earthscan, 2008), p. 3.
15. "National Security and the Threat of Climate Change," The CNA Corporation (Virginia, USA), 2007, p. 6.
16. UNGA Resolution GA/10830, *General Assembly, Expressing Deep Concern, Invites Major United Nations Organs to Intensify Efforts in Addressing Security Implications of Climate Change*, June 03, 2009, see http://www.un.org/News/Press/docs/2009/ga10830.doc.htm, accessed on January 12, 2012.
17. UNSC Resolution SC/9000, *Security Council Holds First-Ever Debate on Impact of Climate Change on Peace, Security, Hearing over 50 Speakers*, Apri 17, 2007, see http://www.un.org/News/Press/docs/2007/sc9000.doc.htm, accessed on January 12, 2012.
18. UNSC Resolution SC/10332, Security Council, in Statement, Says 'Contextual Information' on Possible Security Implications of Climate Change Important when Climate Impacts Drive Conflict, July 20, 2011, see http://www.un.org/News/Press/docs/2011/sc10332.doc.htm, accessed on January 12, 2012.
19. Paul Taylor and Devon Curtis, "The United Nations," in John Baylis, Steve Smith and Patricia Owens (eds.), *The Globalization of World Politics: An Introduction to International Relations 4e* (New York: Oxford University Press, 2008), p. 320.
20. Personal interaction with Cleo Paskal, Associate Fellow, Chatham House, at Manipal University, on September 20, 2011.
21. Personal interaction with Prodipto Ghosh, Member, Prime Minister's Council on Climate Change and Distinguished Fellow, Earth Science and Climate Change at

The Energy and Resources Institute (TERI), on January 03, 2012.

22. Personal interaction with Brajesh Mishra, former National Security Advisor and Principal secretary to the former Prime minister, A. B. Vajpayee, at official residence, New Delhi, on July 07, 2011.
23. Shivshankar Menon, "Non-traditional Security Issues," at 14th Asian Security Conference 2012, Institute for Defence Studies and Analyses, New Delhi, on February 14, 2012.
24. Personal interaction with Group Captain Manoj Kumar, Senior Fellow, Centre for Air Power Studies, in New Delhi, on January 29, 2012.
25. See for more information, "Coral Island," http://www.britannica.com/EBchecked/topic/137072/coral-island, accessed on January 17, 2012.
26. "Sea levels not rising," *The Hindu* (October 15, 2011), see http://www.thehindu.com/sci-tech/energy-and-environment/article2538017.ece?homepage=true, accessed on January 15, 2012.
27. See for more information, "What is Sustainable Development?" http://www.iisd.org/sd/, accessed on January 12, 2012.
28. "Agenda 21," UN Department of Economic and Social Affairs, Division for Sustainable Development, see for more information, http://www.un.org/esa/dsd/agenda21/res_agenda21_01.shtml, accessed on March 18, 2012.
29. Neil Carter, *The Politics of Environment (Second edition)* (New York: Cambridge University Press, 2007), p. 218.
30. John Foster, *The Sustainability Mirage: Illusion & Reality in the Coming War on Climate Change* (London: Earthscan, 2008), p. 10.
31. "The American Clean Energy and Security Act," Centre for Climate and Energy Solutions (Washington, D.C.), see http://www.c2es.org/federal/congress/111/acesa, accessed on January 19, 2012.

Epilogue

This is the way the world ends
This is the way the world ends
This is the way the world ends
Not with a bang but with a whimper.

— T. S. Eliot, 'The Hollow Men'

It is a hard-bitten fact that writings and theories can never be timeless. The theories that were promulgated before the Second World War did not hold good after it, which is why liberalism and idealism gave way to realism and neo-realism. These theories became relatively irrelevant when the curtains were drawn on the Cold War, and a wholly fresh set of theories emerged that could explain the post-Cold War reality. The post 9/11 scenario has thrown fresh challenges to the theorists in terms of explaining and understanding international relations. The world and the individuals who lead international politics and relations keep changing. Every day a new development takes place that renders the previous writings redundant and ancient. However, to dismiss a work on the basis of the era belonged to is naïve as the world might have changed but the geopolitical reality remains the same. The theories that emerged much later were a response to the previous ones. Facts may no longer be the same, but the behavioural patterns and strategies of nation states undergo very little change with the changing times.

In the global environmental debate, environment has always been the loser along with the hundreds and millions of people that live off this environment and its vast resources. Before the Durban Summit (COP-17), the Kyoto Protocol's future was very bleak. Has this changed after the Durban Summit despite the fact that the Kyoto Protocol has been extended by the international community which happened in the last minute of the negotiations as part of a bargain that itself has left several questions unanswered? The CDM and the JI continue to pump blood into the global carbon market that comes at the expense of adaptation, finance and technology transfer; these mechanisms remain but the LDCs and the SIDS continue to wait for the pledged financial and technological resources that would help them adapt to climate change.

No work ends at a dead stop, be it open-ended or close-ended; it is always up to the reader to interpret the work in his/her perspective and give meaning to the work. The last section of any work leaves a void that an author expects the reader to fill as no written work is complete without the reader's interpretation of it as well as his/her response to it. Every written work is an endless chronicle that one needs to go back to in order to begin the narrative as without the beginning, the end can never be reached. No timeless works can be written unless, as the post-modernists say, the power remains with the writer to construe the world in his/her own geopolitical imagination.

An epilogue allows the author to talk directly to the reader in fiction. In non-fiction, it allows the author to bring closure to the work as far as the author's contribution is concerned, leaving the readers to follow the vectors and decide the future. In this section, the focus will be on the Durban Summit that occurred after most of the chapters were written. It is not in the interest of either the author or the reader if some of the calculations made in the chapters have missed the point of exactness by small or big margins. Therefore, the next few paragraphs would outline the latest developments in the global climate change debate with special emphasis on the Durban Summit, and how it could affect the future climate change regime that might or might not unfold in 2020.

The Story—Before, During and After the Durban Summit

Some words reverberate old stories and spark new ones that use the same words but in a different light. An epilogue does not just tell a story; it creates a new one that is continued from the existing one and lays the foundation for the new one. The post-Copenhagen scenario had left several questions unanswered despite a so-called 'successful' Cancun Agreement. The EU has made considerable efforts to address the issue of climate change in the past (while being championed initially and marginalised later); with its historical tradition of propagating universalisation of the European ideals, it took the lead as expected at the Durban Summit (COP-17) to leave its mark as the only 'responsible' power in the world in the midst of economic and social crises. The agenda items at the Summit included:

- debating the future of the Kyoto Protocol.
- operationalisation of several decisions that came out of the Cancun negotiations in 2010.
- Green Climate Fund (GCF).
- mitigation, adaptation, finance, technology transfer, review and legal options.
- a future climate treaty.

The Durban Summit was the longest UN Framework Convention on Climate Change (UNFCCC) meeting ever as two more days were added to the original schedule to salvage the talks. In the Ad-Hoc Working Group on Kyoto Protocol (AWG-KP), the EU declared on the second day of the negotiations that a legally binding framework has to be completed by 2015, while G-77 stuck to its demand for public funds on the part of developed countries for long-term finance, and that the current level of emissions reductions by developed countries was far lower than the expected levels. The LDCs also expressed their concerns over the control of the GCF. On the second day, Mexico came up with a proposal to include a 'last resort' vote should consensus fail on issues that have broad support. This scenario has been looked into in the fourth chapter and the consensus-based approach has officially not been replaced by any other model till now. Russia also pushed for a periodic review of Annex I and II countries but this was opposed by Saudi Arabia. The outcome of the Durban Summit is commonly described as the Durban Package. It consists of:

- the Durban Platform for Enhanced Action.
- the decision on a second commitment period for the Kyoto Protocol to commence from 2013.
- operationalising the Green Climate Fund.
- operationalising the Technology Mechanism.
- operationalising the Adaptation Committee.

The Summit came up with the 'Durban Platform for Enhanced Action', which entails commitments by every country to cut Green House Gas (GHG) emissions; it envisages signing of a deal by 2015 and coming into effect of the same by 2020. Some would call it a success while others would label it 'too little too late'. This document implies that all countries including the US, India and China would have to adopt legally-binding targets; the fact that all countries including those that are not currently part of the Kyoto Protocol have been integrated with the new structure is being viewed by the EU as a great success. The document raises several important questions that need to be resolved in order to call the Summit a real success. It says that the process would 'raise the level of ambition'and 'launch a work plan on enhancing mitigation ambition'.[1] By raising the level of ambition, the platform makes sure that every Party, including developing countries, would have to adopt mitigation measures without adhering to the 'common but differentiated responsibilities and respective capabilities' principle. The wall between the developed and developing countries has been finally brought down. Yet, it is not sure if all the parties are required to take up mitigation obligations as has been seen in the past with respect to the Kyoto Protocol. It is very clear from the text that future commitments by Parties would completely focus on mitigation in the context of 'raising the level of ambition' with adaptation being mentioned only once in the whole document. Even finance and technology development and transfer have been ignored to a large extent. This is in direct violation of the Bali Action Plan that had mandated an outcome which would give equal importance to the four pillars of climate change negotiations—mitigation, adaptation, finance and technology – without which consensus is difficult to arrive at. The wording of the deal's legality is another area that has left several loopholes for the major emitters to escape the law. The weakest description—an 'outcome with legal force' instead of a 'legal instrument'[2]—makes the agreement weak right at the outset.

As far as the GCF is concerned, it is interesting to note that there is great disparity in terms of the funds allocated for mitigation projects vis-a-vis adaptation projects presently. By stating that 'the funding will come from a wide variety of sources, public and private, bilateral and multilateral, including alternative sources of finance',[3] the Copenhagen Accord left a huge question mark over the contribution of each developed country and the credibility of sources. This ambiguity was carried forward from the Copenhagen and the Cancun Accords to the Durban Package too. Thus, the only agreement that was reached at Durban was in terms of setting up a body to distribute and manage the funds, with decisions on sources and allocation still pending. The decision to appoint the World Bank as the interim trustee of the Fund (a decision arrived at the Copenhagen Summit) has not gone down well with many developing countries. The question remains as to how long this interim period could last albeit currently it is stipulated as three years, considering the fact that the international community needs to overcome innumerable hassles to ensure that the pledged amount is garnered and distributed equitably. At the Durban Summit, certain countries opposed the move to make the World Bank the trustee while others tried to reduce the burden on them by calling for a greater private role. So far, Germany has pledged €40 million and Denmark €15 million.[4] Yet another factor raised by several sceptics is the lack of cohesion among developing countries. Some of the developing countries such as the LDCs and the SIDS clearly side by the agenda of the EU, which renders the point of equal representation of the developed and developing countries weaker. The text clearly states that the GCF would be governed and supervised by a Board that would comprise 24 members, with equal numbers form the developed and developing countries. Members from developing countries would include representatives from UN regional groupings, SIDS and LDCs.

The Adaptation Committee's link to the Adaptation Fund is not very clear, and its relationship with the UNFCCC is also shrouded in mystery. Furthermore, there were also allegations that many of the objections and questions of developing countries were brushed under the carpet by the Chairs of the Ad Hoc Working Group on Long-Term Cooperative Action (AWG-LCA) and the AWG-KP (officials from the US and New Zealand respectively) and by the Conference of Parties (COP) President.[5]

The decision regarding the Kyoto Protocol also came under heavy scrutiny by developing countries due to their expectations and the intent expressed by developed countries as there was clearly conflict of interests and beliefs as far as the implementation of the second commitment period of the Kyoto Protocol was concerned. The uncertainty about the Kyoto Protocol got resolved as the EU and a few other Annex-I countries decided to adopt the second commitment period of the Kyoto Protocol, the first commitment period of which ends in 2012. However, the Kyoto Protocol's extension was used as a bargain by the EU to bring the BASIC countries on board for the Durban Platform for Enhanced Action. The question that one needs to ask is – was this not already a legal obligation under the Kyoto Protocol? When the Kyoto Protocol was negotiated upon and brought into force way back in 2005, Annex I countries had also agreed to sign up for the second commitment period when the first commitment period expired. So, to call the trade-off feasible by India and China is highly contestable. On the other hand, it was well articulated by developed countries that the second commitment period was not in their interest in the wake of the economic meltdown and the emergence of global competitors such as China and India. Since the Kyoto Protocol failed to get the stamp of the then largest emitter in the world (the US) and the compliance mechanism failed to bring those countries to book that did not adopt the second commitment period, the compliance procedure of a future treaty needs to be prospective and loophole-free. Canada has refused to sign it and it has already withdrawn from it. Canada would have surely not been able to meet its target of the first commitment period and if it signs on to the second one, it would have to pay penalties under the 'flexible' compliance mechanism of the Kyoto Protocol. The US would still be at par with developing countries and would only be fulfilling voluntary targets during the time between 2012 and 2020. In 2009, President Barack Obama offered a US target for reducing GHG emissions in the range of 17 per cent below 2005 levels by 2020.[6] Also, it cannot be confirmed that countries would eventually reach a legally-binding agreement by 2015, the negotiations for which would begin in 2012.

Role of India

Four things that India demanded are—a second Kyoto commitment from the EU, a GCF, deeper voluntary targets and a delay to the start of talks over a new deal. The Minister of Environment and Forests, Jayanthi Natarajan openly declared that India was forced to sign up or else face the blame for the collapse of the deal.[7] Jairam Ramesh's concessions at Cancun had not paid off since developed countries still stuck to the lowest common denominator although the G-77's demand of adoption of the second commitment period was accepted. India's demands to include unilateral trade measures, Intellectual Property Rights and equitable access to sustainable development (backed by the BASIC, Arab League and the G-77) were not met. The biggest setback was non-inclusion of the term 'equity' in the document. The entire debate came down to the wording of the document on the question of whether to include the phrase 'legal outcome' (as India and China wanted) or a 'protocol or legal instrument' (as the EU wanted) and eventually the phrase—'agreed outcome with legal force'—was endorsed.[8] India even questioned the timing of the demand for a new global treaty citing the fact that developed countries had failed to fulfil the existing obligations.

India lost the plot because of reasons including the reality that the interests of the majority of countries lay in an early global climate agreement while India opposed any move that would impose binding targets on it (albeit it was in favour of a global agreement) alienating it from the rest of the international community. For nearly two weeks, India tried its best to delay the agreement by showing its black flag to any move in this direction. In the end, it was left with no choice but to assent to the Durban Platform without fighting to include principles such as 'equity' and 'common but differentiated responsibility' (CBDR) that form the core of its strategy at the climate change negotiations. The biggest letdown was exemplified by India's obstructive tactics that isolated it from the other developing countries whose interests are heavily tied with that of India. Instead of sounding like a champion of developing country concerns, India appeared as a mere cantankerous irritant that came on the way of an international agreement that could 'save the world'. India's good old strategies of referring to the long forgotten principles did not go down well with any of the Parties.

India has already submitted its own interpretation of the Durban Platform to the UN in which it has included all the factors that are left out of the text such as adaptation, technology transfer, CBDR and others. Since the language of the text is so vague and does not mention emissions reductions or cuts anywhere, India also has the liberty to interpret the phrase, 'highest possible mitigation efforts by all parties'. This also means that the text leaves a huge question mark over the future mitigation efforts by countries and that India is not obliged to adopt binding commitments in 2020 as indicated by the Environment Minister after the Summit.[9]

The Long-Forgotten Goals

Once upon a time the international community was steadfast in its belief that global warming was not happening until it was finally struck with repeated blows of the vagaries of climate change. Some visionaries claimed that global warming was a good thing until scientists proved otherwise. Then the whole debate shifted to whom to pin the blame on—humans or nature itself? Years of scepticism delayed an international consensus until the common goals and objectives were identified when the scepticism slowly began to wane. When consensus was reached as to the international community had to now gear up to take some actions to mitigate the problem, the question of costs and benefits assumed significance. To add to this chaos, the question of who should take up the responsibility of disentangling the problem also complicated the international climate change regime. The negotiations started and they went on and on until a ray of hope pierced through the wall of denial in the form of the Framework Convention and thereafter, the Kyoto Protocol. One fine day (when the talks for the next biggest treaty was supposed to begin), the most vulnerable countries made the clarion call to the same international community that efforts were unsatisfactory, and listed out a few demands:

- Long-term stabilisation of atmospheric greenhouse gas concentrations at well below 350 ppm CO_2-equivalent levels.
- Global average surface temperature increases to be limited to well below 1.5° C above pre-industrial levels.
- Global greenhouse gas emissions to peak by 2015 and decline thereafter.
- Reductions in global greenhouse gas emissions by more than 85 per cent below 1990 levels by 2050.
- Annex I parties to the UNFCCC to reduce their collective GHG emissions by more than 45 per cent below 1990 levels by 2020, and

more than 95 per cent below 1990 levels by 2050, given their historical responsibility.

- A significant deviation from business as usual by developing countries through measurable, reportable and verifiable nationally appropriate mitigation actions in the context of sustainable development supported and enabled by technology, financing and capacity-building, in a measurable, reportable and verifiable manner.[10]

Roger Fisher and William Ury argue in their work, 'Getting to Yes: Negotiating Agreement without Giving In' that principles bargaining produces better agreements than positional bargaining. When it comes to principles bargaining, they outlined four principles—(1) separate the people from the problem; (2) focus on interests rather than positions; (3) generate a variety of options before settling on an agreement, and (4) insist that the agreement be based on objective criteria.[11] These principles are easier said than done in the current geopolitical reality which is driven by positions and exclusive interests. Climate change negotiations have been based on positional bargaining in which each nation state has a different position on every issue and 'the parties bargain from their separate opening positions to agree on one position'.[12] Even though climate change is a global concern and it affects the interests of every nation state, it has been reduced to a problem that needs to be tackled by taking positions that are tagged to the nation state's domestic and foreign policies that delays any action. It has been noted that the positions of nation states are so strong at the climate change negotiations that even if a variety of options are generated, they are struck down due to differences in positions.

Not one of the above-mentioned demands is expected to be met, and the future would continue to hang in balance. If the negotiations were based on science, all these demands would have been priorities for the international community. If the negotiations were based on economics as some experts would claim, externalities would be counted in the cost-benefit analysis. If the negotiations were based on politics, the very credibility of leadership would have been taken into account. If the negotiations were based on values and beliefs, human life would have been given utmost importance.

If the negotiations were based on culture or philosophy, the core of human-nature-society systemic relationship would not remain untouched. In short, the very foundation of the negotiations is shaky. The world requires leaders who can lay down a fresh foundation that can withstand any kind of pressure and at the same time encompass all the facets mentioned above, whether it is economics or culture. If India can pave the way forward by showing different forms of leadership that it has failed to do in the past but has the potential to do so in future, the global environmental order could be repositioned in such a manner that environmental change is not taken as a mere buzzword but an issue that needs to be given utmost priority.

Notes

1. United Nations Framework Convention on Climate Change, Draft decision -/CP.17, "Establishment of an Ad Hoc Working Group on the Durban Platform for Enhanced Action," see http://unfccc.int/files/meetings/durban_nov_2011/decisions/application/pdf/cop17_durbanplatform.pdf, accessed on February 28, 2012.
2. Ibid.
3. United Nations Framework Convention on Climate Change, FCCC/CP/2009/L.7, "Copenhagen Accord," December 18, 2009, see http://unfccc.int/resource/docs/2009/cop15/eng/l07.pdf, accessed on March 19, 2011.
4. Shyam Saran, "India's Climate Change Strategy after Durban," Nehru Memorial Museum and Library (New Delhi), January 06, 2012.
5. Personal interaction with Prodipto Ghosh, Member, Prime Minister's Council on Climate Change and Distinguished Fellow, Earth Science and Climate Change at The Energy and Resources Institute (TERI), on January 03, 2012.
6. "President Obama Sets a Target for Cutting U.S. Greenhouse Gas Emissions," U. S. Department of Energy, December 02, 2009, see http://apps1.eere.energy.gov/news/news_detail.cfm/news_id=15650, accessed on March 19, 2012.
7. "FACTBOX-Winners and Losers of U. N. Climate Deal," *Reuters* (London), December 13, 2011, see http://in.reuters.com/article/2011/12/12/climate-idINDEE7BB0JX20111212, accessed on March 13, 2012.
8. Louise Gray, "Durban climate change: the agreement explained," *The Telegraph* (London), December 11, 2011, see http://www.telegraph.co.uk/earth/environment/climatechange/8949099/Durban-climate-change-the-agreement-explained.html, accessed on March 15, 2012.
9. Chetan Chauhan, "No Emission Cuts Agreed in Durban, says Natarajan," *Hindustan Times* (New Delhi), December 16, 2011, see http://www.hindustantimes.com/India-

news/NewDelhi/India-says-no-emission-cuts-agreed-in-Durban-others-disagree/Article1-783347.aspx, accessed on March 12, 2012.

10. "Alliance of Small Island States (AOSIS) Declaration on Climate Change 2009," see for more information, http://www.aosis.info/documents/AOSISSummitDeclarationSept21FINAL.pdf, accessed on March 15, 2012.
11. Roger Fisher and William Ury, "Getting to Yes: Negotiating Agreement without Giving In," Conflict Research Consortium Book Summary by Tanya Glaser, see http://www.colorado.edu/conflict/peace/example/fish7513.htm, accessed on March 12, 2012.
12. Ibid.

Appendices

- Appendix-A gives a timeline of all the important international meetings that have taken place on the issue of climate change.
- Appendix-B provides a timeline of all the significant events that have shaped the policy and debate of the climate change negotiations.
- Appendix-C is the resolution adopted by the United Nations General Assembly (UNGA) on "Climate change and its possible security implications" on June 11, 2009. This resolution gives an idea of the agenda and objectives of introduction of the issue of climate change in the United Nations Security Council (UNSC). It also illustrates the myriad of interests (of the international community) at stake as far as securitisation of climate change is concerned. The UNSC meetings have been analysed in the seventh chapter in detail.

  Source:http://daccess-dds-ny.un.org/doc/UNDOC/GEN/N08/487/65/PDF/N0848765.pdf?OpenElement, accessed on February 12, 2012.
- Appendix-D is with respect to the "Establishment of an Ad Hoc Working Group on the Durban Platform for Enhanced Action"—an agreement reached at the Durban Summit in 2011. The epilogue emphasises on the Durban Summit and its outcome. The original document could be used to better understand the agreement that has been touted as the biggest breakthrough since the Kyoto Protocol in the climate change negotiations.

  Source: http://unfccc.int/files/meetings/durban_nov_2011/decisions/application/pdf/cop17_durbanplatform.pdf, accessed on December 28, 2011.
- Appendix-E deals with "Increasing Ambition Level under Durban Platform for Enhanced Actions," a submission by the Government of India to the United Nations Framework Convention on Climate Change (UNFCCC). This document essentially puts forth India's stance on climate change and related issues besides its future strategy at the negotiations.

  Source:http://unfccc.int/files/documentation/submissions_from_parties/adp/application/pdf/adp_india_28022012.pdf, accessed on March 13, 2012.
- Appendix-F is the "Joint Statement at the conclusion of the 10th BASIC Ministerial Meeting on Climate Change." The BASIC group of countries have played a more than significant role in the climate change negotiations since the Copenhagen Summit. It is imperative to highlight the official statement of the grouping made during February 13-14, 2012 in New Delhi, India. It yet again throws light on the basic differences between emerging countries and industrialized world as well as the need for the grouping to win the confidence of the international community.

  Source:http://moef.nic.in/downloads/public-information/10th-BASIC-Meeting-Delhi-Joint-Statement.pdf, accessed on March 20, 2012.

Appendix – A

Timeline of International Meetings on Climate Change

Year	Meeting	Venue
1979 (February)	First World Climate Conference	Geneva, Switzerland
1985 (October)	Villach Conference	Villach, Austria
1988 (June)	Toronto Conference	Toronto, Canada
1989 (February)	Ottawa Conference	Ottawa, Canada
1989 (February)	Tata Conference	New Delhi, India
1989 (March)	Hague Conference and Declaration	The Hague, the Netherlands
1989 (November)	Noordwijk Ministerial Conference	Noordwijk, the Netherlands
1989 (December)	Cairo Compact	Cairo, Egypt
1990 (May)	Bergen Conference	Bergen, Norway
1990 (October-November)	Second World Climate Conference	Geneva, Switzerland
1992 (June)	United nations Conference on Environment and Development	Rio De Janeiro, Brazil
1995 (April)	First Conference of Parties (COP-1)	Berlin, Germany
1996 (July)	COP-2	Geneva, Switzerland
1997 (December)	COP-3	Kyoto, Japan
1998 (November)	COP-4	Buenos Aires, Argentina
1999 (November)	COP-5	Bonn, Germany
2000 (November)	COP-6	The Hague, the Netherlands
2001 (July)	COP-6 (Part II)	Bonn, Germany
2001 (October-November)	COP-7	Marrakesh, Morocco
2002 (October-November)	COP-8	New Delhi, India
2003 (December)	COP-9	Milan, Italy
2004 (December)	COP-10	Buenos Aires, Argentina
2005 (November-December)	COP-11/First Meeting of the Parties to the Kyoto Protocol (CMP-1)	Montreal, Canada
2006 (November)	COP-12/CMP-2	Nairobi, Kenya
2007 (December)	COP-13/CMP-3	Bali, Indonesia
2008 (December)	COP-14/CMP-4	Poznan, Poland
2009 (August-September)	World Climate Conference-3	Geneva, Switzerland
2009 (December)	COP-15/CMP-5	Copenhagen, Denmark

2010 (November-December)	COP-16/CMP-6	Cancun, Mexico
2011 (November-December)	COP-17/CMP-7	Durban, South Africa
*2012 (November-December)	*COP-18/CMP-8	*Qatar

* Yet to happen.

Appendix – B

Important Milestones in International Climate Change Policy

Year	Milestone
1987	Release of the Brundtland Commission Report, "Our Common Future"
1988	Establishment of the Intergovernmental Panel on Climate Change (IPCC) by the United Nations General Assembly (UNGA) and the World Meteorological Organization (WMO)
1989	Passing of a UNGA Resolution calling for a global summit on environment and development issues
1990	Publishing of the First Assessment Report of the IPCC
1990	Establishment of an Intergovernmental Negotiating Committee (INC) to draft a framework convention through UNGA Resolution
1992	Adoption of the Rio Declaration on Environment and Development and the United Nations Framework Convention on Climate Change (UNFCCC) by the INC
1994	Coming into force of the UNFCCC
1995	Adoption of the Berlin Mandate at the First Conference of Parties (COP-1)
1995	Publishing of the Second Assessment Report of the IPCC
1996	Adoption of the Geneva Ministerial Declaration at COP-2
1997	Adoption of the Kyoto Protocol at COP-3
1998	Opening of the Kyoto Protocol for signature at the UN headquarters in New York
1998	Adoption of the Buenos Aires Plan of action at COP-4
2001	Publishing of the Third Assessment report of the IPCC
2001	Adoption of the Bonn Agreements at COP-6 (Part II)
2001	Adoption of the Marrakesh Accords at COP-7
2002	Adoption of the Delhi Ministerial Declaration at COP-8
2005	Entry into force of the Kyoto Protocol; first Meeting of the Parties to the Kyoto Protocol (CMP-1)
2007	Publishing of the Fourth Assessment Report of the IPCC
2007	Adoption of the Bali Action Plan at COP-13/CMP-3
2009	Adoption of the Copenhagen Accord at COP-15/CMP-5
2010	Adoption of the Cancun Agreements at COP-16/CMP-6
2011	Adoption of the Durban Platform at COP-17/CMP-7

Appendix C

United Nations A/RES/63/281

General Assembly

Distr.: General
11 June 2009

Sixty-third session
Agenda item 107

Resolution adopted by the General Assembly

[*without reference to a Main Committee (A/63/L.8/Rev.1 and Add.1)*]

63/281. Climate change and its possible security implications

The General Assembly,

Recalling its resolution 63/32 of 26 November 2008 and other resolutions relating to the protection of the global climate for present and future generations of mankind,

Recalling also Article 1 of the Charter of the United Nations, which established the purposes of the United Nations,

Recognizing the respective responsibilities of the principal organs of the United Nations, including the primary responsibility for the maintenance of international peace and security conferred upon the Security Council and the responsibility for sustainable development issues, including climate change, conferred upon the General Assembly and the Economic and Social Council,

Noting the open debate in the Security Council on "Energy, security and climate" held on 17 April 2007,[1]

Reaffirming that the United Nations Framework Convention on Climate Change[2] is the key instrument for addressing climate change,

---

1 1 Including the letter dated 5 April 2007 from the Permanent Representative of the United Kingdom of Great Britain and Northern Ireland to the United Nations addressed to the President of the Security Council (S/2007/186), the letter dated 12 April 2007 from the Chargé d'affaires a.i. of the Permanent Mission of Cuba to the United Nations on behalf of the Movement of Non-Aligned Countries, addressed to the President of the Security Council (S/2007/203) and the letter dated 16 April 2007 from the Permanent Representative of Pakistan to the United Nations, on behalf of the Group of 77 and China, addressed to the President of the Security Council (S/2007/211). See S/PV.5663.

2. United Nations, *Treaty Series*, vol. 1771, no. 30822.

Recalling the provisions of the United Nations Framework Convention on Climate Change, including the acknowledgement that the global nature of climate change calls for the widest possible cooperation by all countries and their participation in an effective and appropriate international response, in accordance with their common but differentiated responsibilities and respective capabilities and their social and economic conditions,

Reaffirming the Programme of Action for the Sustainable Development of Small Island Developing States,[3] the Mauritius Declaration[4] and the Mauritius Strategy for the Further Implementation of the Programme of Action for the Sustainable Development of Small Island Developing States,[5]

Recalling the 2005 World Summit Outcome,[6]

Deeply concerned that the adverse impacts of climate change, including sea - level rise, could have possible security implications,

1. *Invites* the relevant organs of the United Nations, as appropriate and within their respective mandates, to intensify their efforts in considering and addressing climate change, including its possible security implications;
2. *Requests* the Secretary-General to submit a comprehensive report to the General Assembly at its sixty-fourth session on the possible security implications of climate change, based on the views of the Member States and relevant regional and international organizations.

85th plenary meeting
3 June 2009

3. *Report of the Global Conference on the Sustainable Development of Small Island Developing States,Bridgetown, Barbados, 25 April-6 May 1994* (United Nations publication, Sales No. E.94.I.18 and corrigenda), chap. I, resolution 1, annex II.
4. *Report of the International Meeting to Review the Implementation of the Programme of Action for the Sustainable Development of Small Island Developing States, Port Louis, Mauritius, 10-14 January 2005* (United Nations publication, Sales No. E.05.II.A.4 and corrigendum), chap. I, resolution 1, annex I.
5. Ibid., annex II.
6. See resolution 60/1.

Appendix D

Advance unedited version

Draft decision -/CP.17

Establishment of an Ad Hoc Working Group on the Durban Platform for Enhanced Action

Proposal by the President

The Conference of the Parties,

Recognizing that climate change represents an urgent and potentially irreversible threat to human societies and the planet and thus requires to be urgently addressed by all Parties, and acknowledging that the global nature of climate change calls for the widest possible cooperation by all countries and their participation in an effective and appropriate international response, with a view to accelerating the reduction of global greenhouse gas emissions,

Noting with grave concern the significant gap between the aggregate effect of Partiesí mitigation pledges in terms of global annual emissions of greenhouse gases by 2020 and aggregate emission pathways consistent with having a likely chance of holding the increase in global average temperature below 2 °C or 1.5 °C above pre-industrial levels,

Recognizing that fulfilling the ultimate objective of the Convention will require strengthening the multilateral, rules-based regime under the Convention,

Noting decision X/CMP.7 [Title],

Also noting decision X/CP.17 [Title],

1. *Decides* to extend the Ad Hoc Working Group on Long-term Cooperative Action under the Convention for one year in order for it to continue its work and reach the agreed outcome pursuant to decision 1/CP.13 (Bali Action Plan) through decisions adopted by the sixteenth, seventeenth and eighteenth sessions of the Conference of the Parties, at which time the Ad Hoc Working Group on Long-term Cooperative Action under the Convention shall be terminated;
2. *Also decides* to launch a process to develop a protocol, another legal instrument or an agreed outcome with legal force under the Convention applicable to all Parties, through a subsidiary body under the Convention hereby established and to be known as the Ad Hoc Working Group on the Durban Platform for Enhanced Action;
3. *Further decides* that the Ad Hoc Working Group on the Durban Platform for

Enhanced Action; shall start its work as a matter of urgency in the first half of 2012 and shall report to future sessions of the Conference of the Parties on the progress of its work;

4. *Decides* that the Ad Hoc Working Group on the Durban Platform for Enhanced Action shall complete its work as early as possible but no later than 2015 in order to adopt this protocol, legal instrument or agreed outcome with legal force at the twenty-first session of the Conference of the Parties and for it to come into effect and be implemented from 2020;
5. *Also decides* that the Ad Hoc Working Group on the Durban Platform for Enhanced Action shall plan its work in the first half of 2012, including, inter alia, on mitigation, adaptation, finance, technology development and transfer, transparency of action, and support and capacity-building, drawing upon submissions from Parties and relevant technical, social and economic information and expertise;
6. *Further decides* that the process shall raise the level of ambition and shall be informed, inter alia, by the Fifth Assessment Report of the Intergovernmental Panel on Climate Change, the outcomes of the 2013ñ2015 review and the work of the subsidiary bodies;
7. *Decides* to launch a workplan on enhancing mitigation ambition to identify and to explore options for a range of actions that can close the ambition gap with a view to ensuring the highest possible mitigation efforts by all Parties;
8. *Requests* Parties and observer organizations to submit by 28 February 2012 their views on options and ways for further increasing the level of ambition and decides to hold an in-session workshop at the first negotiating session in 2012 to consider options and ways for increasing ambition and possible further actions.

Appendix E

Submission by Government of India
February 28, 2012

Increasing Ambition Level under Durban Platform for Enhanced Actions

1. The CoP 17 decided at Durban to launch a process to develop a protocol, another legal instrument or an agreed outcome with legal force under the Convention applicable to all Parties through an Ad Hoc Working Group on the Durban Platform for Enhanced Action. The CoP 17, inter alia, requested the Parties to submit their views by February 28, 2012 on ëoptions and ways for further increasing the level of ambition and possible further actionsí. India welcomes this opportunity. \
2. As the outcome of the process under Durban platform is to be applicable to all Parties and this outcome is to be reached under the Convention, all principles and provisions of the Convention will apply in totality to both the process and its results. Accordingly, the efforts for increase in the level of ambition must be made, inter alia, in accordance with the principle of equity and the principle of Common But Differentiated Responsibilities of the Parties.
3. India believes that question of ëthe highest possible mitigation efforts by all partiesí has to be addressed in the light of the decisions of CoP 16 at Cancun on ëEnhanced action on mitigationí relating to ënationally appropriate mitigation commitments or actions by developed country partiesí and ënationally appropriate mitigations actions by developing country partiesí.
4. The above CoP decisions on mitigation taken at Cancun recognize that developing country Parties are already contributing and will continue to contribute to a global mitigation effort in accordance with the principles and provisions of the Convention and could enhance their mitigation actions, depending upon provision of finance, technology and capacity building support by developed countries Parties.
5. Developing country Parties including India have already put forward their domestic mitigation goal in accordance with the principles and provisions of the UNFCCC, particularly its article 4, Paragraph 7 which is being implemented accordingly. This can be achieved as envisaged, on provision of support and enablement in terms of finance and technology as per the principles and provisions of the Convention, like Article 4, paragraph 3.
6. Besides, the question of raising the ambition level has to be understood in the context of the paragraph 6 of the decision relating to the establishment of the Durban Platform which states that the process of raising the ambition level will be

informed, inter-alia, by the 5th assessment report of the IPCC, the outcomes of the 2013-2015 review and the work of the subsidiary bodies.

7. Paragraph 5 of the decision on Durban Platform, envisaging the work of increase in ambition level based on the work of the subsidiary bodies is also very relevant. The work of subsidiary bodies includes the work done by the Subsidiary Body on Implementation, Subsidiary Body on Scientific and Technical Advice (SBSTA), as also the two Ad Hoc Working Groups under the Convention and its Kyoto protocol in relation to the various pillars as enumerated in paragraph 5 of the decision.
8. Having regard to the fact that the work plan envisaged in paragraph 7 of the decision on Durban Platform follows the work envisaged/done under paragraph 5 and 6 of the decision, India is of the view that the work plan for enhancing mitigation ambition of Parties relates to the post-2020 period.
9. Considering the fact that the results of the peer-reviewed comprehensive scientific assessments under AR5 and the work of 2013-15 review will be available only in the time frame of 2015, and the work of the AWG-LCA for enhancing the long term cooperation of the Parties under the Convention will not be completed before CoP 18, the only available time frame for making scientific assessment of the ëmitigation efforts by all Partiesí is the post-2020 period.
10. To the extent that the decision, in its preambular recital, makes a reference to the gap between the mitigation pledges until 2020 and emissions pathways consistent with achieving the 2°C goal, the question of ambition in the time frame of 2012-2020 under the Durban platform relates to the commitments of Annex I Parties who have an obligation to take deep and ambitious emission reduction targets consistent with science and the principles of equity and CBDR.
11. It is notable that the paragraphs 7 and 8 of the decision (relating to the Durban Platform) are derived from the text under negotiation in the AWG-LCA relating to the mitigation commitments of the Annex I parties. These paragraphs are now part of the decision on Durban Platform applicable to all parties and have been placed sequentially after the paragraphs 5 and 6, whose import is described above. This confirms the conclusion that the increase in ambition level in the short term is based on the actions of the Annex I country Parties.
12. As the work plan relating to the increase in ambition level mentioned under the Durban Platform is directed at closing the ambition gap with a view to ensuring the highest possible ëmitigation efforts,í and not mitigation results, the increase in ambition levels of Annex-I Parties should be consistent with the findings of science, and the corresponding levels of emission reduction targets needed to stabilize the climate, especially in accordance with the principles of equity and CBDR.
13. In view of the fact that the only accepted scientific basis for determining the level of ambitious actions in the short term is the AR4, the work of the Annex

I Parties for increasing the level of ambition in the time frame of 2012-2020 should be informed by the IPCCC assessments under AR4. AR4 has recommended that Annex I Parties should reduce their emissions at least by of 25-40 percent in the short term by 2020. As the current pledges made by Annex I Parties fall short of the required emission reduction levels, they should raise, in the minimum, their ambition to the level indicated by the AR4. This should be indicated timely and a decision taken thereupon at Doha (CoP 18).

14. Further, the increase in ambition level can be achieved only if the Annex I countries clarify that their commitments/targets in the time-frame of 2012-2020 are without any conditions. This is important to ensure that the pledged actions are clear, ambitious and effective.

15. To facilitate the process of assessing the implementation of commitments of Annex-I Parties towards closing the mitigation gap in the time frame of 2012-2020 in accordance with the findings of science, and the principles of equity and CBDR, a review of the adequacy of the mitigation commitments during this time frame should be launched in accordance with Article 4, paragraph 2(d) of the Convention. The review should be carried out in the light of the best available scientific information and assessment on climate change and its impact as well as relevant technical social and economic information. The results of such review should be an essential part of the 2013-2015 review of implementation of the Convention. The assessed gap in the efforts as compared with the results/ assessment of the Annex-I Parties under Article 4, paragraph 2(d) should be the basis for increasing the level of ambition.

16. In case of Annex-I Parties that are Party to KP, a parallel process should be initiated to determine compliance with their commitments to achieve the economy wide emission reduction targets. The assessed gap in achieving their targets under the first commitment period should be added to their targets in light of above submissions, under the relevant rules applicable to KP Parties. This should apply to all developed country Kyoto Protocol Parties including those that have announced that they will not participate in the second commitment period of the Kyoto Protocol. Similarly, non-KP Parties would also need to increase their targets/actions in light of the review/assessment of their commitments/targets.

17. The Secretariat, in its message vide ODES/CoP17/11 of January 20, 2012 has indicated April 16, 2012 as the date for submission of views of Parties on ëwork plan of the Ad Hoc Working Group on Enhanced Actions including, inter alia, on mitigation, adaptation, finance, technology development and transfer, transparency of actions and support, and capacity building. Indiaís submission thereto will contain further elaboration and/or clarification of the submissions made above.

Appendix F

Joint Statement Issued at the Conclusion of The 10th Basic Ministerial Meeting on Climate Change

February 13-14, 2012 New Delhi, India

1. The 10th BASIC Ministerial Meeting on Climate Change was held in New Delhi on February 13-14, 2012. H.E. Ms. Jayanthi Natarajan, Minister of Environment & Forests of India, H.E. Mr. Xie Zhenhua, Vice Chairman, National Development and Reform Commission of China, H.E. Mr. Francisco Gaetani, Deputy Minister of Environment of Brazil, Mr. Alfred James Wills, Chief Climate Change Negotiator of South Africa and Ambassador Mxakato-Diseko as representative of COP-17 President participated in the meeting. In line with 'BASIC-Plus' approach, Qatar (as incoming President of COP-18), Swaziland (as Chair of Africa Group of negotiators and as a member of LDCs) and Singapore (as member of AoSIS) were invited and participated in the meeting as observers. Algeria (as Chair of G-77 & China) was also invited.
2. BASIC Ministers appreciated the outcome of the Durban Conference and the role played by South Africa as Presidency of COP-17. Ministers recognized that the Durban conference represented a significant step forward and helped operationalize several of the Cancun decisions such as Green Climate Fund (GCF), Adaptation Committee, Technology Executive Committee and the Climate Technology Centre and Networks (CTCN), the Standing Committee on Finance and the arrangements for transparency.
3. Ministers welcomed in particular, the agreement on the 2nd commitment period of the Kyoto Protocol, and stressed that the presentation by May 2012 by Annex-I parties of information on their economy wide quantified emission reduction objectives (QELROs) with a view to adopting an amendment to Annex-B of Kyoto Protocol is an important and necessary first step for the success of the process agreed to at Durban. Ministers reiterated that the flexible mechanisms of the Kyoto Protocol would be available to only those Annex-I parties that have established quantified emissions reduction commitments in the 2nd commitment period. Ministers also emphasized that the non-KP Annex-I parties too must undertake comparable commitments under internationally agreed rules of accounting, measurement, reporting, verification and compliance.
4. Ministers regretted the announcement by Canada, within a few days after the conclusion of the Durban Conference, withdrawing from the Kyoto Protocol.

Ministers observed that the Kyoto Protocol is not only a cornerstone of the international climate regime but a legally binding agreement under the UNFCCC and that any attempts by developed countries to casually set aside their existing legal commitments while calling for a new legally binding agreement seriously questions their credibility and sincerity in responding to the climate crisis. Page 1 of 3

5. Ministers recognized that the Durban Platform offers a clear opportunity for an equitable, inclusive, effective and strengthened climate change regime. Ministers emphasized that the agreement on the Durban Platform was part of a carefully balanced package of 'mutual reassurances' between the parties. They reaffirmed that the process launched at Durban is not to renegotiate or rewrite the Convention and that this process and its outcome shall be under the Convention and in full accordance with all its principles and provisions, in particular the principles of equity and common but differentiated responsibilities and respective capabilities. Ministers welcomed the fact that a compromise was reached at the last minute at Durban to develop a protocol, another legal instrument or an agreed outcome with legal force under the Convention. Ministers noted that the scope of work of Durban Platform has to be defined in advance of the conclusion of the work of the two AWGs.
6. Ministers stressed that unresolved issues such as equity, trade and technology-related IPRs etc must not fall-off the table and remain part of the negotiations.
7. Ministers noted that developing countries are fully committed to playing their part in the global fight against climate change and have presented actions which express significant ambition to reduce emissions. Ministers stressed that developed countries must rise up to their historical responsibilities and take the lead in the fight against climate change by undertaking robust and ambitious mitigation commitments consistent with science and in accordance with the principles of equity and common but differentiated responsibilities and respective capabilities under the Convention.
8. Ministers stressed that equity is a cornerstone of the international efforts at combating climate change and welcomed the decision at Durban to organize a workshop on 'Equitable Access to Sustainable Development'. They stressed that equity must remain an essential element of the work moving forward in the UNFCCC process.
9. Ministers reiterated the importance of the Review of the implementation of the Convention, in accordance with its principles and provisions. Ministers stressed that the clear mandate provided by 1/CP.16 in this regard must be respected. Ministers reaffirmed the important role of the findings of the fifth Assessment Report (AR-5) of the Inter-Governmental Panel on Climate Change (IPCC) in informing the process of implementation of the various decisions of COP-17.

10. Ministers welcomed the operationalization of the Green Climate Fund and called for its early capitalization. They urged developed countries to honor their commitments to provide US$ 30 billion as fast start funding and US$ 100 billion per year by 2020. They also stressed the urgency of securing long-term finance for developing countries including for implementation of Adaptation Framework, national adaptation planning and REDD+. Ministers welcomed the setting up of a platform for discussions of long-term finance under the UNFCCC.
11. Ministers noted with deep concern and reiterated their firm opposition to the inclusion of international aviation in the European Union Emissions Trading Scheme (EU-ETS) which violates international law including the principles and provisions of UNFCCC and runs counter to multilateralism. Ministers noted that the unilateral action by EU in the name of climate change was taken despite strong international opposition and would seriously jeopardize the international efforts to combat climate change. The Ministers recognized the threat of similar unilateral measures being considered by developed countries in the name of climate change in the area of international shipping and expressed their concern.
12. Ministers recalled the first ever statement made by China on behalf of the BASIC countries at Durban and pledged to continue and deepen their cooperation and coordination in the discussions leading up to COP-18 at Doha. In view of the ongoing negotiations for the Rio+20 Conference on sustainable development, Ministers agreed that the BASIC countries should enhance their discussions on Rio+20 issues as well.
13. Ministers emphasized that BASIC countries as part of G-77 & China are extremely vulnerable to the adverse impacts of climate change and therefore share the deep concern of SIDS, LDCs and Africa. The Ministers reaffirmed the need to maintain and strengthen the unity of G-77 & China as the unified voice of developing countries in the climate change negotiations.
14. Ministers welcomed the offer by South Africa to host the 11th BASIC Ministerial Meeting on climate change in the second quarter of 2012.

Index